C0-ABX-574

FORSYTH LIBRARY - FHSU
97___M912n1973
main   Between land

2 1765 0002 7419 4

WITHDRAWN

DATE DUE			

BETWEEN LAND AND WATER

Between Land and Water

The Subsistence Ecology of the Miskito Indians, Eastern Nicaragua

BERNARD NIETSCHMANN

Department of Geography
University of Michigan
Ann Arbor, Michigan

SEMINAR PRESS   New York and London   1973

970.3
M912m
1973

COPYRIGHT © 1973, BY SEMINAR PRESS, INC.
ALL RIGHTS RESERVED.
NO PART OF THIS PUBLICATION MAY BE REPRODUCED OR
TRANSMITTED IN ANY FORM OR BY ANY MEANS, ELECTRONIC
OR MECHANICAL, INCLUDING PHOTOCOPY, RECORDING, OR ANY
INFORMATION STORAGE AND RETRIEVAL SYSTEM, WITHOUT
PERMISSION IN WRITING FROM THE PUBLISHER.

SEMINAR PRESS, INC.
111 Fifth Avenue, New York, New York 10003

United Kingdom Edition published by
SEMINAR PRESS LIMITED
24/28 Oval Road, London NW1

LIBRARY OF CONGRESS CATALOG CARD NUMBER: 72-7703

PRINTED IN THE UNITED STATES OF AMERICA

CONTENTS

PREFACE

My two favorite kinds of places in this world are coral reefs and rain forests. I don't know how I would vote if I had to choose between them, had to decide that I would go only to reefs and never to forests again, or vice versa. My idea has long been to live by a broad, sandy beach with a rain forest behind me and a coral reef offshore before me, with either open to exploration or contemplation. Maybe someday I'll achieve the ideal. There are places where it is possible—some of the islands of the South Seas, or Trinidad and Tobago in the West Indies, for instance.

Certainly they are different enough, the rain forest and the coral reef. They have no inhabitants in common, nor even any general kinds of inhabitants in common. There is no way of comparing their appearance, either. The reef world is bright with color and movement. The forest is all green and brown, dim and still. The reef is Baroque, the forest, Gothic.

—Marston Bates, in THE FOREST AND THE SEA (p. 79)

There is another place where rain forest and coral reefs are close together: the Miskito Coast of Nicaragua. During two research periods, 1968–1969 and summer 1971, I studied the Miskito Indians of Tasbapauni, a village between land and water. During this time I often enjoyed the beauty of these two diverse realms, accompanying Miskito hunters into the rain forest, paddling down sun-flecked forest creeks in dugouts,

ix

and journeying offshore with Miskito turtlemen to sail over coral heads, reefs, and undersea carpets of turtle grass.

I went to the east coast of Nicaragua to study the ecology of Miskito subsistence and how a particular population had adapted to local ecosystems and modified them. The Miskito are one of the few groups left in Middle America for whom hunting and fishing remain extremely important subsistence activities. Much of this study is directed toward explaining the importance of hunting and fishing in Miskito life and subsistence, and analyzing how changes in exploitation pressure of fish and game populations are affecting subsistence, society, and cultural adaptation.

Prefaces are such wooden devices for acknowledging help, advice, and financial support. It is difficult to adequately give thanks for the hours of concern and involvement spent by informants, friends, and advisors during both field research and writing of the manuscript. To all of the following individuals and organizations I wish to express my thanks.

The study was originally completed as a dissertation in the Department of Geography at the University of Wisconsin. William Denevan, Clarence Olmstead, and Henry Sterling from geography, John Hitchcock from anthropology, and Edwin Wallace from the School of Medicine made valuable suggestions for improvement of the manuscript. William Denevan, chairman of the doctoral committee, gave a great deal of stimulation, advice, and editorial help on the original manuscript. In addition, the manuscript was read by John R. Garton of Santa Paula, California, a writer and personal guru of long standing. Roland Bergman, fellow geographer and foraging companion, offered much encouragement and advice during the writing of the manuscript. Miss Heather Coleman was able to decipher my scrawls on yellow pads and type up the rewritten manuscript, which says much for perseverance and cryptography.

The Foreign Area Fellowship Program funded the first 14 months of research and supported writing of the dissertation. The Graduate School at the University of Wisconsin provided funds for processing some of the research materials, maps, and photographs. The second field research period, 4 months in 1971, was financed by a grant from the Social Science Research Council.

Many people in Nicaragua contributed to this study. Ingeniero Cristóbal Rugama, Departamento Geográfico Nacional, made available maps and charts of the coast and the facilities of his office and personnel. Sr. Edgar Curtis, also of the Departamento Geográfico Nacional, who has field checked many of the new 1:50,000 maps of the east coast, provided much information on the present-day distribution of the Miskito. Sr. Jaime Incer of the DGN was extremely helpful in providing materials and facilities during our visits to Managua. Reverend John Giesler, then of the Moravian Church, Bluefields, shared his personal library and knowledge of the Miskito Coast and of the Miskito. Similarly, Reverend David McCallum, Episcopal Church, who lives in the study village of Tasbapauni, provided transportation, suggestions and amenities of the outside world which were greatly appreciated. Doctor Edwin Wallace, then of the Gray Memorial Hospital in Puerto Cabezas and Doctor Peter Haupert of the Thaeler Memorial Hospital in Bilwaskarma, gave much of their time, experience, facilities, and advice, as did their families. M. J. D. Hancock, Forestry Officer, FAO Mission to Nicaragua, flew me along the coast to take photographs of Miskito settlements and exploitation areas. Richard Bishop, a Peace Corps volunteer working on a fisheries project in the Bluefields area, helped identify fish and other little beasties and shared his small cache of fine teas. Brian Weiss and Marianne Schmink, students in the Department of Anthropology, University of Michigan, helped with fieldwork in 1971, and were a welcome diversion for Miskito gringo-watchers.

The Miskito of Tasbapauni were especially tolerant and understanding of my research efforts. Baldwin Garth and his family shared their home and friendship, making fieldwork in the village enjoyable and memorable. Many trips and adventures were made with Mr. Baldwin and other Miskito and their attention to detail and careful explanation of their world taught me a great deal. Mr. Flannery Knight and Cleveland Blandford of Tasbapauni took great care in describing the Miskito cosmos and were especially tolerant of my initial attempts at handling dugout canoes at sea and machetes in the bush. Visiting and working with the Miskito along the coast and rivers were always a joy and an education.

Three other individuals have lived with this study for as long as I have: Judi, my wife, our son Barney, and Sabi, our margay. Judi's contributions extend far beyond field research, photography, and editing, for she was admired for her enthusiasm, insights, and companionship by the Miskito and husband alike. Our son adapted well to life in Bluefields and Miskito villages and to the rigors of field research in the tropics. He became equally comfortable eating green turtle meat and cassava as if they were a hamburger and french fries. Even as I write this, our margay is watching the typewriter keys as they hit the page. He has watched every paragraph being written, almost as if he wanted to be sure that what I said about his former home was correct.

QUOTATION CREDITS

Page ix   Bates, Marston. *The forest and the sea.* Signet Science Library. New York: New American Library, 1960.

Page 3   From Sahlins, Marshall D. Notes on the original affluent society. In *Man the hunter,* edited by R. B. Lee and I. Devore. Pp. 85–89. Chicago: Aldine, 1968.

Page 4   From Clarke, William C. *Place and people: An ecology of a New Guinean Community.* Originally published in 1971 by The University of California Press; reprinted by permission of The Regents of the University of California.

Page 5   From Mikesell, Marvin W. Cultural ecology. In *Focus on geography,* edited by Philip Bacon. Pp. 39–61. Washington: National Council for the Social Sciences, 1970.

Page 6   From Vayda, Andrew P. and Rappaport, Roy A. Ecology: Cultural and non-cultural. In *Introduction to cultural anthropology,* edited by J. A. Clifton. Boston: Houghton Mifflin Company.

Page 7   From Rappaport, Roy A. *Pigs for the ancestors: Ritual in the ecology of a New Guinean people.* New Haven: Yale Univ. Press, 1968.

Pages 36 and 73   From Carr, Archie. *The windward road: Adventures of a naturalist on remote Caribbean shores.* New York: Alfred A. Knopf, Inc., 1956.

Page 57   From Helms, Mary W. *Asang: Adaptations to culture contact in a Miskito village.* Gainesville, Florida: Univ. of Florida Press, 1971.

Pages 63 and 88   From Klingel, Gilbert C. *The ocean island (Inagua).* A Doubleday Anchor book, Natural History Library. Garden City, New York: American Museum of Natural History, 1961.

Page 80   From *The cloud forest* by Peter Matthiessen. Copyright © 1961 by Peter Matthiessen. Reprinted by permission of The Viking Press, Inc.

Pages 87–88   From Parsons, James J. The Miskito pine savanna of Nicaragua and Honduras, *Annals of the Association of American Geographers* **45**, 36–63.

Page 98   From Coe, Michael D. and Flannery, Kent V. Early cultures and human ecology in south coastal Guatemala. *Smithsonian Contributions to Anthropology* **3**. Washington: Smithsonian Institution.

Page 100 (footnote)   From Porter, Philip W. Environmental potentials and economic opportunities—a background for cultural adaptation. Reproduced by permission of the American Anthropological Association from the *American Anthropologist* **67**: 412, 1965.

Page 103   From Conzemius, Eduard. *Ethnographical Survey of the Miskito and Sumu Indians of Honduras and Nicaragua.* Smithsonian Institution, U.S. American Ethnology Bulletin 106, Washington: Smithsonian Institution, 1932.

Page 109   From Woodburn, James. An introduction to Hazda ecology. In *Man the hunter,* edited by R. B. Lee and I. Devore. Pp. 49–55. Chicago, Aldine, 1968.

Page 123   From Chisholm, Michael. *Rural settlement and land use: An essay in location.* London: Hutchinson, 1962.

Page 132   From Leeds, Anthony. Yaruro incipient tropical forest horticulture: Possibilities and limits. In *The evolution of horticultural systems in South America, cause and consequence: A symposium,* edited by J. Wilbert. Pp. 13–46. Caracas: Sociedad de Ciencias Naturales La Salle, 1961.

Page 163   From Bennett, Charles F. Animal geography in Latin America. In *Geographic research on Latin America,* edited by B. Lentnek, Robert L. Carmin, and T. L. Martinson, Pp. 33–40. Muncie, Indiana: Ball State Univ., 1971.

Page 178   From Matthiessen, Peter. To the Miskito Bank. In *The New Yorker,* October 28, 1970, pp. 120–164.

Page 180   From Carr, Archie. *So exellent a fishe: A natural history of sea turtles.* New York: The Natural History Press, 1967.

Page 182   From Sahlins, Marshall D. *Stone age economics.* Chicago: Aldine, 1972.

CHAPTER I

INTRODUCTION: THE ECOLOGY OF SUBSISTENCE

Large numbers of societies live at the "subsistence level" in tropical Latin America. A variety of subsistence systems characterize diverse indigenous groups. Many of these systems already have been greatly modified by external, more highly organized societies; some are presently undergoing large-scale change, and a few represent little altered vestiges of once widespread systems of subsistence.

Tropical rain forests and tropical inshore waters are two of the most complex, stable, and biologically productive ecosystems on the Earth. Many indigenous cultures which interact with these ecosystems have adapted their food resource strategies so that ecological integrity is protected. Increasing national and international demand for tropical resources for the development of extraneous systems is, however, creating an unstable ecological and economic situation and is increasingly disrupting traditional subsistence systems. The vastness, stability, and diversity of tropical ecosystems have historically acted to resist the intrusion of more complex industrialized societies upon less complex, subsistence societies. With intensifying external need for materials, energy sources, and land for pioneer settlement and monocrop agriculture, the circle is closing; rain forests are being cut back, tropical forest peoples are being—or already have been—acculturated, assimilated, depleted, or

1

exterminated. Self-sufficient, internally regulated subsistence systems are being replaced by externally dependent and controlled monetary market systems. The great variety and diversity of cultural and ecological systems once found in the tropics have been simplified, disrupted, and weakened.[1]

In this study of the coastal Miskito of eastern Nicaragua, the subsistence system of a human population is the major concern. The study focuses on the interrelationships of human subsistence needs and means and environmental stability within a local system; and how forces generated from larger and more complex social and economic systems have changed, disrupted, and are destroying the ecological and social stability of the Miskito system.

Several refuge areas survive where shifting agriculture, hunting, fishing, and gathering still form the basis of subsistence for indigenous groups. In Oceania, Africa, and Latin America, geographers and anthropologists especially have been focusing on ecological relationships between subsistence-based peoples and their environments.[2] It is from the vantage point of tropical subsistence peoples that we are learning much about the relationship of subsistence systems[3] to population patterns, ecological stability, food productivity, and the rational utilization of tropical ecosystems.

Until very recent ecological research, our interpretation of subsistence peoples has usually explained more about how we think than about the societies investigated. The study of subsistence systems has suffered

---

1. Rappaport (1971) presents a detailed argument for this viewpoint.

2. For example, see Clarke (1971), Denevan (1971), Harris (1971), Rappaport (1968, 1971), and Waddell (1972). Brookfield (1968) and Nietschmann (1971b) provide survey articles on recent research on tropical agriculture and subsistence in the tropics.

3. A subsistence system can be thought of as the complex of functionally related resources and activities through which a group secures food for its own needs and by its own efforts, usually by the direct exploitation of its environment. The primary objective is food, whether it is from agriculture, horticulture, silviculture, hunting, fishing, gathering, or animal husbandry. Production, distribution, and consumption of foodstuffs are generally performed by discrete social units, such as a household or kin group, with little circulation of labor or produce outside the social network. Subsistence groups make their own living rather than earning it. In some societies, production is for consumption and consumption is by the producers, while in other societies, some domestic production may be for exchange, with return receipt of foodstuffs. In subsistence economies, production intensity is mainly geared to the food needs of the producers with possible surplus production for religious and social purposes, for livestock, or because of reoccurring environmental hazards.

The degree of reliance on production for consumption varies depending on involvement with external exchange or monetary market relationships. Various provisioning relationships could be placed along a hypothetical continuum ranging from "pure" subsistence systems to market systems in terms of decreasing production for direct consumption and increasing purchase or exchange of foods. To state it another way, production–consumption relationships are regulated mostly by internal homeostatic mechanisms in closed subsistence systems and are increasingly regulated by external mechanisms in open or market systems.

from two basic weaknesses: the lack of detailed, quantitative data for more meaningful analysis, and the obstructing economic philosophy of our Galbraithean-based explorations into the economic life of primitives and peasants. The very word "subsistence" conjures up images of a hard, marginal life, continuous work just to survive, inability to produce surplus, low return from labor, little security of life, poor diet and nutrition, and a universal level of livelihood which is an impediment to economic development (Nietschmann 1971b). Examples of these subsistence stereotypes in textbooks and articles are too numerous to mention and are generally incorrect. We are often told that indigenous systems of swidden agriculture and hunting and fishing are not only unproductive but also biologically wasteful and ecologically disruptive.

The prevalent view of subsistence economies and peoples will be difficult to counter and rebuild without a reconsideration of our economic preconceptions. With respect to hunters and gatherers, Sahlins (1968:85) remarked that "perhaps then we should phrase the necessary revisions in the most shocking terms possible: that this was, when you come to think of it, the original affluent society." As Sahlins, notes, there are two roads to affluence: by either satisfying wants through producing much, or by desiring little. The assumption in our economic system is that man's wants are great and his means limited. For many primitive peoples, however, wants are limited and means are great. Or, to put it another way, market economies are based on inadequacy and deprivation while subsistence economies are based on adequacy and dispensation.

> Inadequacy is the judgement decreed by our economy, and thus the axiom of economics: the application of scarce means against alternate ends. We stand sentenced to life at hard labor. It is from this anxious vantage that we look back on the hunter. But if modern man, with all his technical advantages, still hasn't got the wherewithal, what chance has this naked savage with his puny bow and arrow? Having equipped the hunter with bourgeois impulses and Paleolithic tools, we judge his situation hopeless in advance [Sahlins 1968:86].

Very little is known about the content, structure, or ecology of subsistence. Information and analysis are lacking on the productivity relationships between labor, technology, and resources; the ability of subsistence systems to provide sustenance to populations; the factors of subsistence leading to long-term ecological stability; and the dietary quality of the food produced. In general, the scale of subsistence has not been accounted for, nor has subsistence been counted with a scale. Ongoing research is suggesting the need and means to reevaluate the productivity, reliability, adaptability, and capability of subsistence systems.

Subsistence behavior and system organization may be viewed

ecologically; subsistence is a major interface between interacting human and environmental components of an ecosystem. The main point of interchange between a subsistence-based culture and its environment is in the procurement of food. Subsistence economies and strategies are forms of cultural adaptation, that is, systems by which man adapts to his environment in order to maintain a viable relationship. In so doing, subsistence strategies may be adjusted and environments modified to assure a group's survival and well-being within tolerable limits of the cultural system and the ecosystem.

The use of an ecological viewpoint and a systems approach in the study of subsistence departs from most earlier research in geography and anthropology which was hampered by treating either cultures or environments as idiosyncratic isolates or opposing forces that had deterministic influences on each other. Rejecting deterministic and possibilistic controls, several geographers and anthropologists have been focusing on relationships within ecosystems as their object of analysis. Neither the environment nor a human population or its culture are viewed as acting on the other in a unilinear way, but are considered parts of an interacting system which, through its circular relationships and systems of negative and positive feedback, influences and modifies each one, and changes them together (Clarke 1971: ix).

> As it is now applied to human life by anthropologists and geographers, the idea of the ecosystem, which stresses the circularity of the relationship between organism and environment, makes it easier to consider that environment as both a result of and an influence on human behavior. Man affects the environment; in turn, the changed environment requires new responses from man and acts to rearrange man's image of his surroundings. Considered thus as components of an ecosystem, both man and his environment are seen as parts of a single unit, the whole of which is worthy of study. Concern shifts from which part most influences the other to the structure of the whole system and how it operates and changes [Clarke 1971:200].

An ecosystem is considered "a set of living organisms and nonliving substances interacting to produce an exchange of material between the living and non-living parts [Odum 1959:10]." The concept of the ecosystem encourages an emphasis on obligatory relationships, interdependencies and causal relationships between living organisms and their environment, both physical and biological. The ecosystem includes "all of the organisms . . . in a given area interacting with the physical environment so that a flow of energy leads to clearly defined trophic structure, biotic diversity, and material cycles (i.e. exchange of materials between living and nonliving parts) within the system [Odum 1971:8]."

Ecosystems are so complex that only subsystems or particular relationships may be examined with any hope of comprehension. The ecological relationships described in this study, for example, are but a small set of variables and elements from the total ecosystem. As Mikesell (1970) noted,

> the concept of a system embracing the totality of nature and culture is probably beyond the grasp of any one scholar or any one scholarly discipline. Consequently, the most common strategy . . . is to examine a particular link in a particular ecosystem, the connection, for example, between climate and agriculture. Establishment of the functional character of a link of this sort permits expansion into other components of a system, e.g., technology, social organization, landforms, and soils [p. 42].

Food-getting activities and their adaptation to and impact on environment form major links between human subsistence populations and diverse cultural, physical, and biological components in the ecosystem. By studying the ecology of subsistence within an ecological matrix, some of the functional relationships which couple and regulate man–environment interchanges may be identified and measured. In this way one may be able to achieve a more realistic analysis of the interaction between human populations and their environments, rather than simply concluding that subsistence peoples lead a precarious life in a never-ending struggle against Nature.

Cultural systems and ecosystems can be thought of as organized patterns for the processing of information, energy, and materials. The transfer of energy and materials from the ecosystem to the human system is primarily through culturally guided patterns of resource evaluation and exploitation. A subsistence system, then, is the assemblage of technologies and strategies with which humans modify and exploit energy relationships in order to tap and control biotic systems in the supply of energy and materials for human sustenance and maintenance.

In order to maintain a long-term flowthrough of energy and materials, exploitation pressure has to be regulated and balanced to permit the ecosystem to maintain its stability and regenerative capacity. If the relationship between human populations and the ecosystem can be regulated through some cultural mechanism so that negative feedback results, then stability may be maintained. If, on the other hand, energy and material requirements increase in the human system, then positive feedback occurs, and a new relationship must be achieved between the culture and the ecosystem.

A constant shifting and reshuffling of variables and self-regulating elements adjust and adapt human populations to an environment. Whether or not a steady state or homeostasis is attained, or even exists in human ecosystems, is questioned. Clarke (1971) observed that "cultural behavior may act to counterbalance environmental changes, but there is always a net change in the system. Certainly, ecosystems as evolutionary entities are self-maintaining, but they are self-transforming too. Rather than a homeostat, a gyroscope may be the suitable analogy for regulating mechanisms in ecosystems [p.202]." What is important in maintaining a stable relationship between man and environment is that negative feedback mechanisms are effective in identifying variables deviating from ecologically acceptable conditions and returning them to tolerable levels. If changes are more rapid than the system's ability to detect and react satisfactorily in time, then equilibrium relationships may be broken, creating a deviating, amplifying condition. Thus, it is the system's capacity to detect change and the speed at which change occurs that are important in system regulation.

Subsistence behavior and food procurement strategies can be thought of as adaptive mechanisms between human populations and the environment. A subsistence system is regarded here as a set of relationships which regulates a population's interaction with its environment. Thus, a group must obtain food if it is to survive, and an environment must not be overexploited if it is to survive in a nondegraded form. Between a group's biological need for food and an environment's capacity to supply it, is the regulatory mechanism of a subsistence system.

It is necessary to make several very specific statements about the theoretical and methodological bases upon which this study is based. First, the focus is upon human populations and their cultural adaptation[4] to a set of varied environments for subsistence, and the impact of this adaptation upon the biotic communities. Human populations are viewed as being another component in the ecosystem; thus, they are

> commensurable with the other units which they interact to form food webs, biotic communities, and ecosystems. Their capture of energy from and exchanges of material with these other units can be measured and then described in quantitative terms. No such advantage of commensurability obtains if cultures are made the units, for cultures, unlike human populations, are not fed upon by predators, limited by food supplies, or debilitated by disease [Vayda and Rappaport 1968:494].

What survive beyond human–environment interchanges and crises are cultural traits and adaptations, an understanding of which is the final goal. In the analysis of human population as an interacting component in an ecosystem, definite methodological and conceptual advantages

---

4. For the purposes of this study, culture is considered a form of adaptation of human populations, which is exemplified by behavior patterns, energy flowthrough patterns, cultural artifacts, and information systems. Culture is the adaptive link which unites humans with their environment and serves as a blueprint for the creation and maintenance of the human habitat.

are obtained in terms of measurement and in the application of ecological theories; all of which have the ultimate aim of explaining cultural patterns and their interrelationship with the environment, and how these patterns change and readapt. If we can measure and understand the functional mechanisms or sets of variables which adapt a society to an environment, in this case the subsistence system, it may be possible to identify adaptive and maladaptive trends in cultural systems and ecosystems. Systemic alterations in the relationship between human populations and their environments, whether caused by internal or external forces, help to create new cultural adaptations which will serve as new guidelines for human behavior and interaction with environments.

Rappaport (1968) suggested that by analyzing the relationships between human populations and ecosystems, one can make important contributions toward the goal of elucidating cultural phenomena:

> A population may be defined as an aggregate of organisms that have in common certain distinctive means for maintaining a set of material relations with the other components of the ecosystem in which they are included. The cultures of human populations, like the behavior characteristics of populations of other species, can be regarded, in some of their aspects, at least, as part of the "distinctive" means employed by the populations in their struggles for survival. It has been suggested by the biologist G.G. Simpson (1962:106) that the study of cultural phenomena within such a general ecological framework may provide additional insights into culture, "for instance, in its adaptive aspects and consequent interaction with natural selection" [p.6].

There are tremendous operational and epistemological problems in identifying the parts, much less the structure and functioning of a system which involves such diverse and complicated components as a human population and a tropical rain forest. The flow of energy is one of the major elements which organizes ecosystems and distinguishes a subsistence system. Many of the rates, patterns, and exchanges involved in energy flow can be identified and measured, whether for a human system or a biotic community.[5] Measuring techniques, however, are often inadequate or equipment is too costly or bulky. It is impossible to measure everything; one is lucky if he can measure and trace a few of the energy paths in human societies and between those societies and the environment. From a small core of measured and observed relationships, one tries to tie together the many loose ends to produce a unifying description and explanation of the overall "system."

5. The results of energy flow measurement in a tropical rain forest in Puerto Rico have been reported in a massive work edited by Odum and Pigeon (1970). Rappaport (1971) described and analyzed energy flow patterns among a group of swidden agriculturalists in the tropical forests of New Guinea.

Field researchers may be able to measure some of the functional elements in a system but have to then attempt to fill in the explanatory interstices, using Western models and classifications and their own cognitive appraisal of what constitutes the "whole." Is the system a viable, real world entity, or only an association of diverse elements over which the researcher has infused a unifying structure? Systems sometimes may be detected by measured relationships but they can also exist only on the pages of a research proposal or report.

"As a partial solution to the uncertainty about what is relevant and operative in a given human ecosystem, some anthropologists have advocated the use of ethnoscience or, more specifically, ethnoecology [Clarke 1971:203]."[6] To learn what a group perceives and how it classifies its environment, is to identify what it considers relevant and worthwhile for human use and activity. Just because an indigenous classification system does not conform to Western notions of reality does not mean that it is without ecological significance. As Vayda and Rappaport (1968) noted "it is reasonable to regard a people's cognition with respect to environmental phenomena as part of the mechanism producing the actual physical behavior through which the people directly effect alterations in their environment [p.490]." As part of the cultural complex with which a group adapts to an environment, ethnoclassification systems are important determinants in the perception and utilization of environmental resources.

Perception of available resources and the nature of exploitative behavior are to a large extent culturally determined. These can be either ecologically adaptive or maladaptive, depending on how feedback information influences behavior, values, and cognition.

There are few human populations left in the world which do not interact with other systems, other human populations. The economic, social, or political intrusion of a more highly organized system on a less organized one creates in the latter a period of intense rearrangement and restructuring of social and ecological relationships. When two systems of disparate size and organizational complexity are articulated, the energetic implications can be described ecologically:

> There is some energy exchange between the two subsystems in the sense that the less-organized subsystem gives energy to the more-organized, and, in the process of exchange, some information in the less-organized is destroyed and some information gained by the already more-organized [Margalef 1968:16].

6. "The aim of the ethno–ecological approach is simply to present a people's view of the environmental setting itself and their view of behavior appropriate to that setting [Vayda and Rappaport 1968:490]."

The sudden flux in the demand for matter or energy on a local system by an external system may bring into conflict the stability and organization of traditional adaptive mechanisms which have evolved over a long period of time. Homeostatic regulations and feedback mechanisms which served to maintain human population size and the level of energy circulation, may be circumvented. Intensity of demand influences the rate of energy and material transfer within a system and from one system to another. If the rate of transfer and change becomes greater than the local system's capacity for self-corrective adjustment and ability to restabilize, a new level of human–environment interchange has to be found. The speed, frequency, and types of energy transfer affect the ecology of subsistence: the organization and scheduling of production, distribution, and consumption. If the subsistence system can no longer regulate the major articulation between human populations and the ecosystem, internal controls may be lost along with ecosystem stability.

This is the same situation which faces the people in this study: attempting to maintain the control of a viable, ecologically conservative subsistence system and at the same time participating in a monetary-based market economy with links to extraneous systems and controls. With competing new values and goals, control is being lost over the regulation of subsistence and severe disruptions are being created in the interchange within and between the population and the environment.

We are considering the Miskito as a population within an ecosystem whose major adaptations are expressed through culturally guided behavior. Their subsistence is regarded as a coherent system with which the Miskito interact with the ecosystem and with each other. It is our thesis that subsistence is an adaptive system, organized and scheduled to reduce subsistence risk and to adjust the human population to the ecology of an area. If either the environment experiences major ecological changes, or the human population size changes, or energy and materials are diverted to extraneous systems, there should be concomitant changes in the subsistence system, adjusting the population to the new situation.

The Miskito's subsistence system is analyzed in terms of its (1) historical development and adaptation, (2) structure and composition, (3) as a regulator of energy flow between and with human populations and the human ecosystem, (4) relationship to and impact on biotic communities, (5) productiveness and dependability, and (6) response to external system demands for energy and materials.

The approach in this study then, focuses on the ecology of a subsistence system as it functions in and on the ecosystem, including, whenever possible, measurement and ethno–ecological explanations of Miskito attitudes and behavior. Through the subsistence system Miskito cultural and behavioral patterns are transmitted to selected ecological effects on their environment, influencing, in turn, succeeding behavior and culture in general. Many of these patterns are expressed definitively in land use, through their impact on populations and associations of native biota, as cultural adjustments to the ecology of resource availability, and in the form and function of the subsistence system itself.

CHAPTER II

FIELDWORK ON THE
MISKITO COAST

If you are in the tropics and have trouble seeing the good in where you are, work your way to windward where the trade comes in to land.

—Archie Carr, in THE WINDWARD ROAD (p. vii)

One has an opportunity to do a lot of thinking while traveling by boat. I recall one trip by dugout from my field base to a Miskito village some 70 miles distant. The constant droning of the outboard motor could not detract from the quiet magnificence of the tree-lined creeks and riverine corridors through the tropical forest. Such beauty draws one to reflection, and as I turned through the twisting channels that led through the muddy shoal waters, by now automatically picking out the right creek route from the many choices, I remembered my first conversation with a Miskito, a man of about 45. He watched me as I came up the path to his village from the lagoon landing where the diesel boat from Bluefields had just tied up.

"How is it?" I asked him, using the Creole phrase for "hello."

"Right here," he answered, eyeing me inquisitively.

"That's good." "Tell me, where can I find the oldest man in the village?"

"Oldest man? Oldest man?" "Oldest man, him dead!"

Here I was, ready to begin fieldwork, to investigate people's activities and behavior, to measure their land, their labor, their harvests, their hunting and fishing, even to measure their food, and the first man I talk to measures me! The image of that moment stayed with me and is one of the rewards of field research with the Miskito.

THE MISKITO COAST

The Miskito Coast[1] stretches along the eastern coastal plains of Honduras and Nicaragua from Cabo Camarón in the north to the Río San Juan in the south. By far the greatest part of the Miskito Coast is in Nicaragua. The coastal lowland of eastern Nicaragua is part of one of the largest coastal plains in Middle America, extending over 600 miles in length from Costa Rica into Honduras (West 1964:81). This littoral environment has a marine zone of shallow offshore waters strewn with coral cays and reefs varying from about 75 to 10 miles in width, plus a coastal lowland extending inland as far as 100 miles.

The vast Atlantic lowland of Nicaragua is scarcely known to the outside world nor even to the majority of Nicaraguans residing in the Pacific region. Parsons (1955a) remarked that the east coast of Nicaragua is one of the "least known, least visited, and most forgotten parts of the entire Caribbean area [p.54]." Similarly, very little scholarly attention has been directed toward the Miskito Indians. Spread along the coast are a number of cultural groups, of which the Miskito are the most numerous, being the second most widely distributed Indian people in Central America. Distribution of the Miskito fits generally within the area of the coastal lowland, except for a few Miskito villages along the upper reaches of the Río Coco, well into the foothills of the mountain backbone of Nicaragua. The Miskito are scattered along 400 miles of Caribbean coast in Nicaragua and Honduras. They extend far up the major rivers that pass through the verdant lowland tropical forests and seemingly endless savannas of the coastal lowland and into the reaches of the higher tropical forest.

Miskito population is concentrated along the middle and lower Río Coco and along the coast from Old Cape, just south of Cape Gracias a Dios, to Haulover, 30 miles north of Bluefields (Fig. 1). In this study

---

1. Following Parsons (1955a) and Helms (1971) I will refer to the east coast of Nicaragua as the Miskito Coast, rather than by the more common "Mosquito Coast." The origin of the term *Mosquito Coast* is quite uncertain. Several hypotheses have been suggested but they are not convincing. Since the Miskito are the largest cultural group and the coast's history has long revolved around the Miskito, it seems correct to call it by its rightful name, the Miskito Coast, home of the Miskito Indians.

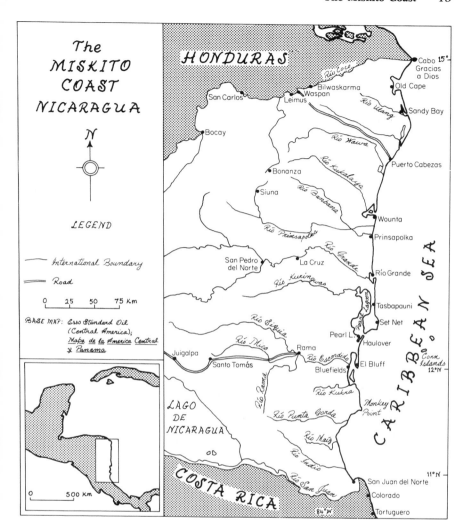

Fig. 1. General location map of the east coast of Nicaragua. For a more detailed map see Fig. 3, p.16.

we will focus on the coastal villages, especially the village of Tasbapauni, with occasional references to riverine Miskito settlements away from the coast.

The coast remains today culturally and geographically isolated from the rest of Nicaragua. There are only very limited means of transportation into the area. Frequent Lanica air flights by rickety C46's go from

Managua to Bluefields, Corn Island, Puerto Cabezas, Waspam, and the mining area in the mountains. A newly completed road extends from Managua to Rama on the Río Escondido, where a boat can be taken to Bluefields. Occasionally, a few hardy individuals will journey to the coast by going down the Río San Juan, Río Grande, Río Prinsapolka, or Río Coco. But these are long and arduous journeys. Local transportation is largely by passenger–freight boats which traffic along the coast, principally between Bluefields and Puerto Cabezas, the two major ports, and up many of the rivers. In the northeast area of the coast there are many old logging roads in various states of disrepair, the major one connecting Puerto Cabezas with Waspam and Bilwaskarma on the Río Coco.

The principal market and port towns are Bluefields (population 16,000) and Puerto Cabezas (population 8000). Ports such as San Juan del Norte (Greytown), Río Grande, and Prinsapolka are no longer important, being mere survivors of past economic booms. Away from the coast, market towns are located at Bilwaskarma, Waspam, Leimus, and San Carlos on the Río Coco and at Bonanza and Siuna in the mining area. Supplies can be purchased in some of the villages along the Río Grande and Río Prinsapolka.

Because of the riverine and coastal distributions of the Miskito, and the nature of the low, swampy littoral environment, broken by large lagoons, and dissected by numerous rivers and creeks, transportation for the Miskito has been and continues to be largely by water or on foot. There is a regular flow of Miskito Indians walking along the beaches and across the northern savannas on their way to buy supplies, to look for work, or to visit distant relatives and friends. Because of the frequent journeys by foot, dugout canoes, and small diesel boats, many of the Miskito intimately know their environment for long distances from their home. Some Miskito have traveled to the very edges of their world and many are equally at home in forest or sea.

In addition to the Miskito there are four other Indian groups presently inhabiting the Miskito Coast: the Rama, and three subtribes of Sumu Indians, including the Ulwa, Panamaka, and Twahka (Fig. 2). There are also two villages of Black Caribs on Pearl Lagoon—Orinoco and La Fe—established in the late 19th century by Black Caribs brought from Honduras to cut mahogany. A large percentage of the population of Bluefields, San Juan del Norte, Corn Island, and Pearl Lagoon are "Creoles," mixed descendants of Negroes, English, and some Indians. The term Creole in eastern Nicaragua can be used for anyone from an almost pure Jamaican Negro to an almost white Anglo descendant from the Bay Islands, Cayman Islands, or San Andrés. Spanish-speaking

Fig. 2. Four of the cultural groups on the east coast of Nicaragua. In the foreground is a Rama, next an Ulwa Sumu, a Miskito, and a "Creole." Photograph taken at the conference of Moravian lay pastors, Tasbapauni, July 1969.

Ladinos from the interior of Nicaragua, or second generation from the coast, make up a sizable segment of the population in Bluefields and Puerto Cabezas, and in towns and villages such as San Juan del Norte (where Spanish-speaking Creoles are called "Black Spaniards"), Prinsapolka, the mining area, and Waspam. Only a very few Ladinos are found in any of the Miskito villages, but recently a rapidly advancing agricultural frontier of Ladino farmers is moving from the interior mountains onto the Miskito Coast. The frontier is so mobile that its location on the map is only approximate (see Fig. 3, p.16).

FIELDWORK AND RESEARCH

My wife and son accompanied me on both research trips to Nicaragua. For the first research study, September 1968 to October 1969, we established a field base in Bluefields, the major town on the coast and a center for communications. After a 2-week reconnaissance of some Miskito villages to the north, I selected the village of Tasbapauni to work in, located 70 miles north of Bluefields through coastal lagoons, rivers,

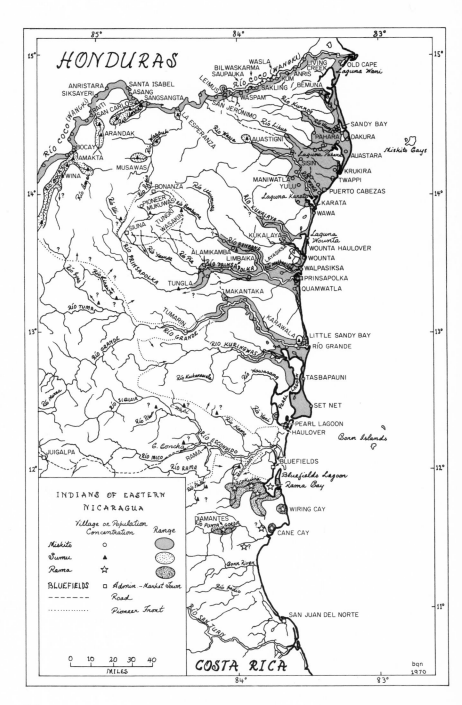

and creeks. It offered most of the things I was looking for in a study site. The people exploited a number of different environments—terrestrial and marine—that were analogous to a large part of the coast; some Creole English[2] was spoken so that the attempt to learn Miskito would not be too difficult; and many other villages were accessible from the field site. Fieldwork was carried out in Tasbapauni for 4- or 5-week periods, with a few days off to return to Bluefields for supplies or to visit other nearby villages. During our second field study, May through September 1971, we stayed in Tasbapauni for the entire period.

The study village of Tasbapauni was founded in about 1860 by Miskito from Sandy Bay. Conditions were said to have been bad in Sandy Bay and the Miskito King, George Augustine Frederick, suggested that some of the people should move. According to the grandson of one of the original settlers, the village was started by George Hayman and Lawrence Julius and their wives. The site was known to the Miskito from their hawksbill expeditions to the San Juan del Norte area. The Tasbapauni location was selected because of its haulover site[3], rich hunting grounds, and most importantly, for the Sandy Bay turtlemen, because of the offshore turtle banks and cays. In the following years additional Miskito came to "Panka Karban" (Rolling Barrel) which later was called Tasbapauni (Red Earth).

Tasbapauni lies on a narrow haulover, separating Pearl Lagoon from the sea (Fig. 4). Approximately 1000 people live in some 200 houses distributed along a half-mile beach front.[4] Houses are pieced together from rough-hewn wood and odd widths of drift lumber found on the beach. Roofs are usually thatched with palm leaves; wealthier individuals use corregated zinc, status figuring as a higher requisite than comfort in the tropical climate. The houses are raised off the ground by stilts of termite-resistant wood. Each house usually contains three rooms, a front porch, and either an inside kitchen or one separate from the house, depending on the economics of the particular family. Roaming cattle, goats, and horses graze on low, thick grass. Pigs and chickens scurry in and out from under houses, bare-bottomed children chasing

2. The English-speaking inhabitants (Jamaican, West Indian, Cayman Islanders, Corn Islanders, and some Miskito) of the Caribbean coast of Nicaragua refer to their dialect as "Creole." For detailed and contrasting views regarding English as a dialect, a Creole, or a Pidgin, see *Pidginization and Creolization of Languages* (Dell Hymes, ed. 1971).

3. A haulover is a narrow neck of land separating two bodies of water such as a lagoon and the sea or two lagoons, where a *dori* can be crossed over.

4. Village census totals were 997 in July 1969, and 1014 in May–June 1971.

Fig. 3. Distribution of the Miskito, Sumu, and Rama Indians, eastern Nicaragua, 1969.

Fig. 4. The village of Tasbapauni, August 1969. View is looking southeast toward Pearl Lagoon, in the background.

them. Abundant coconut palms, breadfruit, and mango trees provide shade.

Three churches representing Moravian, Anglican, and Catholic denominations are located in the village. Each has a resident Miskito lay pastor, and, in the case of the Anglican Church, an American missionary. Three "stores" supply dry goods and clothes brought from Bluefields. A *guardia*, or government soldier, lives in the village and operates a small generator for wireless communication to Bluefields. A government doctor is supposed to visit the village once or twice monthly and sees patients in a two-room clinic built by the people of Tasbapauni with the help of the Nicaraguan government and the Alliance for Progress.

Six diesel-powered boats, built and operated by Tasbapauni men, run to and from Bluefields carrying freight, passengers, an occasional green turtle destined for market, and coconut oil. These boats are made from large dugouts, split down the middle, widened with planks, and decked over. Most have ancient low-horsepower diesel engines, which frequently refuse to run. Two or three boats will usually make the 70-mile one-way trip to Bluefields every week. Travel time varies from 10 to 24 hours, depending on whose boat it is and whether or not it breaks down or runs aground. These boats are the village's major means of contact with other villages and with markets.

The beach is strewn with tangles of logs, driftwood, coconut husks, and the remains of butchered turtles. Under the coconut palms are thatched sheds, jammed with dories, pulled up on bamboo rollers out of the sun. A group of turtlemen may be stretched out in the dories, discussing where the turtle are and whether conditions are good enough to go to sea after them. One of the turtlemen will probably be wedging a piece of old rag into a crack to caulk a leak in his *dori*; or he may be painting over a newly made repair with whatever color is handy. Just outside the *dori* shed turtle shells from the morning's butchering will be boiling in a rusty oil drum over a dried coconut husk fire. As soon as the shell gets soft it will be peeled off and the calipee[5] taken out. Behind the sheds, flour sack sails swing in the wind, their masts stuck in the ground, while young boys splash off dried salt with buckets of fresh water from nearby wells.

The major portion of the study was done in Tasbapauni and with the coastal Miskito. By focusing on one village we were able to collect over a 1-year period detailed data on agriculture, hunting and fishing, and dietary patterns, which provided a datum to evaluate the subsistence system and changes that had occurred in the 1½-year interval between our 1968–1969 and 1971 work.

Comparative research was done in other Miskito villages along the coast and up rivers far into the interior. Travel was largely by *dori*, *pitpan* (two types of dugout canoes), or small, diesel-powered passenger–freight boats reminiscent of the "African Queen." Reconnaissance trips were made along the entire coast from the Río San Juan to the Río Coco, to the offshore islands and cays, and 350 miles up the Río Coco. We had a *dori* with an outboard motor, which greatly facilitated fieldwork, allowing us to visit agricultural and hunting grounds far removed from the villages. In all some 2500 miles were covered in our *dori*, and another 1000 or so in other boats.

Brief research visits were also made among other cultural groups on the coast including Rama and Sumu Indians, Creoles, and Black Caribs. These visits provided a clearer picture of the overall ecology of the area and helped to put Miskito culture patterns in better perspective.

---

5. Calipee is the cartilaginous amber and dark gray-colored substance obtained from inside the lower and upper shells of the green turtle. It is used as the basis for green turtle soup in the United States and England.

In order to get information on how food was produced and used by the Miskito, procedural field outlines by Conklin (1963) and Green-field (1965) were followed. I made up an outline for hunting and fishing, consisting of topics to be covered in each village. Foods were weighed with a dietary scale and their approximate composition was taken from a table by Wu Leung (1961). It was impossible, due to the isolation of these villages, to send food samples to Managua for analysis. The Wu Leung table suffices for many of the foods eaten by the Miskito, but many of their foods are not shown. Here only approximate estimates could be given.

The area which the villagers used for the procurement of food was mapped with the aid of 1 : 40,000 airphotos, 1 : 50,000 topographic maps, 35-mm transparencies taken from a light airplane, and mostly from ground reconnaissance with the aid of Miskito guides. Much of my interest was directed toward exactly where agricultural produce, game, or fish were taken. Miskito guides provided descriptions of each locale and how they categorized it, including the nature of the land and what plants and animals were characteristic.

Daily records were kept during both field periods of the amounts of game meat and fish brought into the principal study village by hunters and fishermen. The weights of these animals were taken both of the field dressed and butchered meat. The exact location at which meat was taken was also recorded, as well as the distance and time involved. This was done in order to get at the meat yields of respective biotopes and the relative success in different biotopes and with different animals, as expressed by amounts taken relative to time and distance expended. The time, distance, yields, and type of animal and biotope provided a core of measurable data.

With this information, along with data on agricultural yields and labor inputs, and the types and approximate amounts of wild fruit collected, as well as the amounts of purchased foods, it was possible to describe productivity ratios of different food-getting systems.

Dietary surveys of three families were done at various times of the year in Tasbapauni. This involved visiting each of the families at mealtime and weighing their food. My wife collected information on how each of the foods was prepared. More dietary work of this nature was to have been done but the surveys involved so much time which detracted from other research that we thought it best to do only short sample studies. Families eat at different times which may change every day. Therefore, we had to stay within sight of the selected families, watch for their cooking fires, and then hurry from one to another. At times, an attempt to weigh some food before a Miskito put it in his mouth was, to say the least, difficult. Also it was considered polite to offer food to the inquisitive, intruding *meriki waikna* (American man). To decline such food would have been an insult. It was very difficult to be a detached observer without at least minimal participation because much of the daily life of the people is centered around food; to step into a kitchen is to be given food.

A typical daily routine involved getting up at 5:30 A.M. for a circuit of the village to record the game animals brought in the night before, and then accompanying a man up the beach or across the lagoon to his "plantation"[6] until late in the morning. Meanwhile, my wife, Judi, visited with women, talking to them at home or working with them in the swiddens. After lunch, field notes were written. From 2:00 to 5:00 P.M. interviews were held with people in their homes. About 5:00 P.M. another walk around the village was made to record game animals and to get an idea of the types and amounts of food brought in from the plantations. Nights were spent listening to stories about "how it used to be" from an old *kuka* or *dama* (grandmother and grandfather), and in attempting to scribble notes in the dark. We returned home early to prepare for the next day's work: a hunting trip or out to sea after turtle. If this was the plan, coffee and bread were readied in advance, then a few hours sleep until a 2:00 A.M. knock and whisper at the window: "De land-breeze is here, we go fo' de turtle."

In our absence from the principal study village of Tasbapauni, informants kept the daily records. Their efforts were usually as accurate as ours and often less time-consuming, as they were tuned in on the "Black Radio" ("jungle telegraph") which could pass along with amazing speed word of meat coming into the village.

Obtaining information, of course, was simply not a matter of being in the village to weigh, measure, and record subsistence efforts and results. One had to be sensitive to the entire social complex, and able to perceive the rationale for Miskito behavior and explanation. For example, a young boy was walking along the beach carrying a few pounds of freshly butchered meat tied by a strip of bark. "From who you get that meat?" I asked in my mixture of broken Miskito and Creole English. "One man," he replied. "And where does that man live?" I queried. "Yonder," he said and walked off. Now "yonder" is not too helpful in the identification and location of someone. But to the boy there was no need to identify and locate, everyone already *knew*. The meat was obviously deer, and the boy was walking from uptown, and everyone knew that most of the men were out after turtle and that Burnell was the only hunter left. Therefore my question was superfluous—"yonder" meant Burnell's.

6. "Plantation" is the commonly used term for a cultivated swidden. In Miskito it is *insla* and in Spanish *milpa*.

CHAPTER III

EXPANSION AND CHANGE

The Miskito were once a small population inhabiting a relatively restricted area in contrast to their current situation. Their emergence as the dominant indigenous people in eastern Nicaragua and their diffusion into very different environments has come about since European contact.[1] Rapid increase in population, territorial expansion, and a long period of culture contact have placed considerable pressure on the Miskito's subsistence system and on their use of the environment. A better understanding of the Miskito's present-day relationship with their environment and the nature of their subsistence system can be gained by looking at the historical and cultural antecedents.

The Miskito have been in long association with Europeans, chiefly English-speaking peoples, and outside market opportunities since at least the early seventeenth century: "Hunting and fishing, gathering of natural resources for barter and sale, and wage labor have all been equally important to the Miskito economy since its origins in the seventeenth century [Helms 1971:4]." For more than 300 years there have been transfers of energy and materials from the Miskito and their ecosystems to outside societies and systems. In return, the Miskito acquired trade goods, money, firearms, and political support, especially from the English.

---

1. Helms (1971) presents a detailed account of the Miskito's postcontact economic history.

Outside economic demand for resources from the Miskito environment was mainly for nonsubsistence items. Contact and trade were intermittent and resource extraction periodic. The rates of resource depletion and Miskito involvement varied according to market demand, resource availability, and other factors, as pointed out by Helms (1971): "The speculative nature of these investments, combined with additional factors such as plant disease, exhaustion of resources, government instability, wars, and depressions tended to produce a rather irregular economy characterized by small booms and subsequent busts [p.27]."

Throughout their long period of intimate market contact with outsiders, it has been mainly the adaptability of the Miskito's traditional indigenous subsistence system which has enabled them to participate in two worlds: their own kin-based, reciprocal exchange, subsistence society, and the foreign wage labor and money market economy. This was possible because for most of the postcontact period, the Miskito maintained control of their subsistence system and their ecosystem, and were able to make important changes in their society and culture to adapt them to the new situation. Nevertheless, the long-term diversion of energy and materials to outside systems has left the Miskito with a much-degraded environment and cultural adaptations to wage and market opportunities which are not well-fitted to the realities of the present-day depressed economic situation on the coast.

Even though much of Miskito culture history is very little known and highly speculative, there being almost no information on the precontact situation and only brief and sporadic accounts during the first years after prolonged contact with Europeans, it is safe to say that postcontact Miskito culture came to be different in many respects from what must have been the precontact aboriginal situation. Many of the cultural adaptations made during three centuries of contact were structural changes in Miskito society, in their rules of behavior and ecological accord with the environment, which began to adjust the Miskito more and more to outside economic forces rather than to maintaining an internal harmony between and within the society and the ecosystem. Nevertheless, it has been the subsistence core of traditional Miskito culture and adaptation which has absorbed the social and ecological strains of economic change and population expansion.

The prolonged close economic relations between the Miskito and Europeans, even though periodic, have been of enough significance to prompt Helms (1971) to claim that the "existence of the Miskito as an identifiable ethnic group with a distinctive way of life is a direct result of trade with the West [p.228]." To be sure, and as Helms documents, there have been many social and economic changes by the Miskito to culture contact, but these have been adaptations—not spontaneous creations.

The Miskito's distinctiveness as a group with a well-defined adaptation to a littoral environment was quite evident at the time of contact with Europeans. The Miskito did not "originate" a new culture or go through a cultural metamorphosis as a result of trade with the West. They did, however, make extensive cultural adaptations to new economic systems and helped to transform and transfer energy and materials from their ecosystem to overseas systems. Therefore, their capacity to regulate the interchange between humans and the environment was reduced, and new cultural patterns became evident.

In this study, we will argue that the Miskito of the early seventeenth century had a well defined, sea-oriented culture with a subsistence system focused on fishing, hunting, and a lowland tropical forest agricultural system; that the Miskito were a distinct group recognized by themselves and outsiders, and that most of their traditional subsistence system has persisted for the more than 300 years since contact.

EARLY SUBSISTENCE AND SETTLEMENT

The aboriginal groups which inhabited eastern Nicaragua prior to contact had cultures and languages closely affiliated with lowland tropical forest culture in South America (Adams 1956: 879–899; Kirchhoff 1948; Helms 1971:14). Characteristic to the east coast and the Caribbean coast of Central America was a complex of South American culture traits including: an emphasis on hunting and fishing; the cultivation of roots and tuber-yielding plants rather than seed crops; the dependence on dugout canoes for travel; and the use of hammocks and bark cloth (Helms 1971:14).

To describe the aboriginal culture it is necessary to turn to early writings of buccaneers and adventurers who provided the first detailed information about the Miskito. From 1502, when the coast was discovered by Columbus, until the early seventeenth century, the area and the Indians were largely ignored by the Spanish. The first Europeans to have friendly relations with the Miskito were French buccaneers who visited Cape Gracias a Dios in about 1612 (Exquemelin 1686, II:277). According to Long (1774:317) the English made contact with the Miskito about 1630. Parsons (1962:30) gave 1633 as the date when a trading station was established at Cape Gracias a Dios with the Miskito by "English adventurers" from Old Providence Island.[2] The first extensive accounts of the Miskito start with A.O. Exquemelin, author of the earliest widely read work on buccaneers, who was on the coast in 1671 or 1672, and

2. Other dates for the first Providence trading settlement at Cape Gracias have been suggested. Parsons (1954:8) cites 1632 and Newton (1914:165) gave 1634 as the date.

William Dampier, the well-known English navigator, who visited Corn Islands, Bluefields Lagoon, and Cape Gracias a Dios in 1681. The French buccaneer, Raveneau de Lussan, and his crew crossed Nicaragua in 1688 from the West coast to the East coast, descended the Río Coco on rafts, and encountered Miskito Indians in a village near Cabo Gracias a Dios. The best description of the seventeenth century Miskito was written in 1699 by an Englishman, known only by his initials, "M.W." These accounts and those from other early writers are the major historical descriptions of the Miskito not long after first prolonged European contact.

Early historical accounts indicate that the Miskito had a small population living in scattered villages distributed along the coast north and south of Cape Gracias a Dios (M.W. 1732:286–288; De Lussan 1930:283–284). The Miskito had a very pronounced subsistence focus on the littoral environment, particularly coastal lagoons and offshore turtle grounds. If, as has been suggested, the aboriginal Miskito were once a coastal subtribe of Sumu,[3] possibly affiliated with the Bawihka (Conzemius 1932:17, Helms 1971:18), then they were the only subtribe with such a distinctive sea orientation. If, on the other hand, the Miskito existed as a group linguistically and culturally apart from the Sumu, then part of their distinctiveness and identity was due to their coastal location and subsistence focus.

The present-day linguistic, cultural, and physical characteristics that distinguish the Miskito from the Sumu may, indeed, not have always been so pronounced. Miskito and Sumu languages are part of the Chibchan linguistic family of northern South America (Greenburg 1960; Stone 1966:210), and, according to Lehmann (1910:714), are closely related structurally but are quite distinct as dialects. Cotheal (1848), in his early analysis of Miskito grammar, noted that Miskito differs "so much from the neighboring tribes that they are unintelligible to each other without the aid of interpreters [p.238]." On the other hand, Helms (1971:19ff) suggests that the distinctiveness of Miskito language could have come about as a result of culture contact. In 1699, M.W. (1732) noted, however, that the Miskito "live peacefully together in several families, yet accounting all Indians of one tongue, to be the same people and friends and are in quality all equal . . .[p.293]."

Due to their coastal location and ready acceptance of foreigners, the Miskito early received biological admixtures of outsiders into their population, starting in the seventeenth century with buccaneers, traders, and

---

3. "Sumu" is a collective name for linguistically and culturally related subtribes which once included Twahka, Panamaka, Ulwa, Kukra, Bawihka, Yusku, Prinsu, Boa, Silam, and Ku (Conzemius 1929a:64; 1932:15; 1938:936). The last seven are now extinct.

shipwrecked Negro slaves, and continuing until the present day. The Sumu Indians did not permit intermarriage and kept aloof or retreated from foreign influence (Conzemius 1932:12–14, 17; Helms 1971:18). Several historical sources mention a slave ship that was wrecked on the Miskito Coast sometime in the 1640s and whose survivors were taken in by the Miskito and accepted into the indigenous population (Conzemius 1932:17; Exquemelin 1856:167; De Lussan 1930:285). The Miskito Indians "are all of a dark yellow or brown complexion, having long black hairs, excepting the *Mullattoes*, whose black hair curls; and their bodies are nearer to the color of negroes, from whose mixture with the *Indians* they first sprung, occasioned 50 years since [1649] by a *Guiney* merchant ship which was driven to leeward, having lost her way, and perished on the coast [M.W. 1732:293]."

The Miskito were a waterside people. Their villages, plantations, transportation, and meat-getting activities were intimately tied to water. In referring to the Miskito north of Cape Gracias a Dios, M.W. (1732) indicated that they "inhabit along the sea-shore, pretty close to the sea-side, or on the sides of some lakes or lagunes [sic] hard by [p.286]." One of the major reasons for their riverine or coastal settlement was to procure meat; especially fish and turtle. According to Dampier (1968):

> There is Land enough, and they may choose what spot they please. They delight to settle near the Sea, or by some River, for the sake of striking Fish, their beloved Employment. . . . After the Man hath cleared a Spot of Land, and hath planted, he seldom minds it afterwards, but leaves the managing of it to his Wife, and he goes out a striking [p. 16].

At early contact, the Miskito were located in small scattered settlements along the northeastern coast of Nicaragua and the adjacent coast of Honduras, primarily around Sandy Bay and Cape Gracias a Dios. De Lussan (1930) remarked in 1688 that "the original inhabitants of Moustique . . . have settled ten or twelve leagues to the windward of Cape Gracias a Dios, at places known as Sambey and Sanibay [p.287]." Their distribution was given by M.W. in 1699 to extend from Cabo Camarón in Honduras to what is today called the Wawa River, south of Puerto Cabezas in Nicaragua (1732:286–288). Settlements were made up of a few houses of related families (M.W. 1732:293), and distributed along the coast, lagoons, and rivers near the sea.

The total Miskito population was probably not a very large one, numbering perhaps less than 2000 in the mid-seventeenth century. Exquemelin (1856) claimed that "they are in all but a small nation, whose number exceed not sixteen or seventeen hundred persons [p.167]." Dampier (1968:15) estimated that in 1681 the male population of Cape Gracias a Dios, a major village, was under 100 and M.W. (1732:287) reported that there were but 400 people in Sandy Bay, the principal Miskito settlement.

The Miskito's expertise in handling their dories in surf and at sea, and their great skill in turtling and fishing so awed the European buccaneers, good sailors in their own right, that almost every account makes frequent mention of Miskito seamen. For example, the much-traveled buccaneer, De Lussan (1930), observed that the Miskito

> are the boldest in the world in braving the perils of the sea and are without dispute the most dextrous in fishing. They go out to sea in small boats that the average sailor would scorn;[4] in these they remain three or four days at a stretch as unconcerned, despite the weather, as if they were part of the boat. Once a fish is sighted, no matter how far under the water, they never fail to get it so great is their skill [p. 286].

Most of the descriptions by Exquemelin, Dampier, De Lussan, and M. W. are fragmentary, but they do suggest that the Miskito population was relatively small, located near water, and that agriculture was not emphasized even though there appeared to be no shortage of land. On the other hand, all writers make frequent references to the Miskito's skill and preoccupation with hunting and fishing. Dampier (1968), for example, in describing the Miskito, reported:

> Sometimes he seeks only for Fish, at other times for Turtle, or Manatee, and whatever he gets he brings home to his Wife, and never stirs out to seek for more till it is all eaten. When hunger begins to bite, he either takes his Canoa and seeks for more Game at Sea, or walks out into the Woods and hunts for Peccary, Warree, each a sort of wild Hog, or Deer; and seldom returns empty-handed, nor seeks for any more so long as any of it lasts [p.16].

Exquemelin (1856) observed that the Miskito were so expert and unfailingly proficient at striking fish, turtle, and manatee that the buccaneers always took them on board on their voyages to help feed the ship's crew: "for one of these Indians is alone able to victual a vessel of one hundred men [p.166]."

The coastal littoral environment was rich in aquatic animals, and much of the Miskito's subsistence came from the sea, coastal lagoons, and rivers. The waters around the Miskito Cays abounded with green turtles; the coastal lagoons were filled with varieties of fish and shellfish and manatee; and the rivers and gallery forests were abundant with wild animal protein sources.

---

4. Sea-going dugout canoes (dories) are radically different from the blunt-ended river canoes (pitpans) used by riverine peoples. The design and manufacture of dories for turtling and fishing by the coastal Miskito is one diagnostic trait which distinguishes them from the upriver peoples whose pitpans would be useless at sea.

Great differences existed between the coastal Indians and those living more in the interior along the river in the means and ease of subsistence. Bell (1862) noted this in the mid-nineteenth century when he observed:

> The Coast Indians drink twice as much as those inland, and at the same time are on the whole much healthier, which can only be accounted for by their superior living: for while the sea and the brackish lagoons afford them abundance of turtle, large sea-fish, and shell-fish, the poor Indian of the interior hooks a scanty meal for his family of small river-fish by much patience and toil, or pursues, often unsuccessfully, the fleet game through the tangled woods [p.261].

The Miskito's agriculture was not mentioned much by the early writers, partly due to the fact that swidden agriculture must have seemed very crude and careless to the foreigners who were used to carefully tilled fields in Europe. Dampier (1968) recorded that the Miskito made a very small plantation which did not seem to him to provide enough for their needs. Their largest plantation contained not more than "20 or 30 Plantain-Trees, a Bed of Yams and Potatoes, a Bush of *Indian Pepper*, and a small Spot of Pine-apples . . . [p.16]." The plantations were small and were located near riversides "at a good distance from their dwelling houses [M.W. 1732:296]." In addition to the cited foods, the Miskito cultivated bananas, sweet cassava, coconuts, pejibaye, sugarcane, papaya, and a little corn in their plantations; they gathered wild honey, hone palm seeds, locust, sapodilla and coco plum fruits, and crabs; and they hunted and fished for an abundance of river and sea fish, deer, white-lipped peccary, and many kinds of birds and turtles (Exquemelin 1856:167; M.W. 1732:293, 296–298).

The combination of hunting and fishing and swidden agriculture provided food in abundance and much leisure. With food so easily secured, much of the Miskito's time, judging from the buccaneer writers, was spent sleeping in hammocks. As might be expected, sleeping during the day went against the Western work and activity ethic of the visitors who made constant references to "lazy" and "idle" Indians. For example, De Lussan (1930) commented in 1688 that

> they are extremely lazy and only plant and cultivate sparingly. They lie all day in hammocks (a kind of swinging bed) in their wigwams or huts, while their wives do their work. Only when pressed by hunger do they embark in their boats and go fishing—for which they have unusual aptitude. When they made a good catch, a feast is prepared and they do not work again until faced by pangs of hunger [p.287].

Even though subsistence was their major preoccupation in terms of time inputs, even this apparently did not take too much effort. As M.W. (1732) observed in 1699: "These people lead a very idle life, not taking any pains, except in hunting, and going to fish in their doreas [sic] or boats made out of a whole piece of wood, and in keeping same in repair [p.293]." He added:

> It is in the morning that they go out to fish or hunt, and what they get they bring home to their wives to dress for them; which victuals may serve them perhaps for two days, and some fruits; during which time the men have no more work to do, but to swing in their hammocks, unless some extraordinary matters of state intervene, or consultations with their *Sukias* [shamans] about invading the Alboawinneys [Sumu], or robbing the Spaniards, or on the notice of being invaded by either of them, and such like [p.293].

EXPANSION

Through their close contact with English and French buccaneers, and English traders, the Miskito soon learned how to use firearms and were known as brave fighters. Dampier (1968) noted:

> When they come among Privateers, they get the use of Guns, and prove very good Marks-Men: they behave themselves very bold in fight, and never seem to flinch nor hang back; for they think that the white Men with whom they are, know better than they do when it is best to fight, and let the disadvantage of their Party be never so great, they will never yield nor give back while any of their Party stand [p. 16].

As soon as they obtained firearms the Miskito began to make raids against the Spanish and neighboring Indians. Warfare had long occurred between the Miskito and different Sumu subtribes but the similarity of traditional weapons and size of fighting parties had apparently kept the situation fairly even (Conzemius 1932:81–82; Dampier 1968:16; Helms 1969a:77–78; M.W. 1732:287–288). Whereas formerly they were not an extreme danger as an enemy because of their small number and scattered distribution, the Miskito, once armed with guns and ammunition, began to extend their territory and influence. Sumu Indians had little direct dealing with the buccaneers and traders and consequently did not obtain guns. M.W. (1732) observed that the Sumu of the Río Waspuk (near the mid-Río Coco) were "extremely terrified at the firing of a gun, out of which, they say, an evil spirit issues [p.290]."

In 1687 a Miskito King was established by the English, who wanted to substantiate their claim to the coast with a government that swore allegiance to them. This gave the Miskito raiding parties a certain legality as punishment was meted out for failure to pay tribute[5] to the Miskito King.

----

5. Tribute articles collected from the Rama Indians, for example, included hawksbill shell, dugout canoes, hammocks, and cotton lines (Roberts 1827:100).

Through the seventeenth and eighteenth centuries the Miskito conquered or drove back all of the Indian tribes from the coastline of Nicaragua. Many groups were forced to retreat to the interior headwaters, usually beyond the rapids, away from the aggressive and militarily superior Miskito. By the late seventeenth century the Miskito had advanced their settlements as far north as the Río Tinto in Honduras, and southward, by the end of the eighteenth century, to the southern end of Pearl Lagoon and up many of the major rivers (Conzemius 1932:83–84; M.W. 1732:285, 291). The Miskito controlled almost the entire Atlantic coastline of Honduras and Nicaragua and continued to push their raiding expeditions in large sea-going dories and river pitpans against Indian and Spanish villages in Honduras, Nicaragua, Costa Rica, and Panama. At the end of the eighteenth century, the Miskito had gained an ascendancy over most of the Caribbean coast of Central America (Conzemius 1932:83–84; Helms 1969a:80; Bell 1862:242; Roberts 1827:49, 71, 86, 199; Roys 1943:70, 120).

The Miskito population grew partly through the inclusion of remnants of defeated peoples and runaways[6] (Nietschmann 1969). The offspring of all outsiders brought in or accepted into Miskito society grew up as Miskito, learning the culture and language of their adopted group. In the seventeenth century, according to M.W. (1732), the Miskito were raiding Sumu Indians and taking away "their young wives and children for slaves, either killing or putting to flight the men and old women [p.291]." The Miskito also captured many Rama and Sumu Indians and sold them to foreigners as slaves. According to Stephen Kemble, who visited the Coast in 1781, "numbers of Ulwa Sumu had been seized on at different times by the Mosquito Men, and sold to vessels trading to Hispaniola, Jamaica, and North America, as well as the British Settlers on this Coast [1884]."

The Miskito often combined voyages and travels for resources with raids on other groups. They made annual expeditions to Costa Rica for hawksbill shell and green turtles to exchange with the English for trade goods. During their journeys, which lasted weeks, sometimes months, they had occasion to attack nearby Indian and Spanish settlements, especially in the Matina Valley of Costa Rica. Pittier (1892) claimed

---

6. As Helms (1971) notes: "In addition to shipwrecked slaves, Negroes escaping servitude on West Indian plantations and in the Spanish mines in interior Honduras probably sought refuge on the isolated Coast. Negro slaves were also brought by seventeenth-century English planters and small settlements of Indians and Negroes grew up, especially at Bluefields and Cape Gracias (Squier 1858:633; Helbig 1959:179; Floyd 1967:21) [p. 16]."

Negroes from Jamaica were brought by English settlers in the eighteenth century and West Indian Negroes emigrated later. The Miskito intermarried with and assimilated Negro males into their villages. Children of Miskito–Negro parents "always speak the language of the mother and grow up as Miskito [Conzemius 1932:13]."

that the Miskito had been raiding against lowland settlements in Costa Rica from the beginning of the seventeenth century:

> They [Miskito] obtained guns and ammunition from Jamaica and on their annual voyages along the coast they would begin plundering and burning. They not only seized all objects of value but destroyed buildings, killed the Spanish and took the Indians in order to sell them as slaves to their protectors in Jamaica. Their piracy had such success that all of the Atlantic Coast remained depopulated during the eighteenth century; the crops abandoned and maritime trade declined [p.112].[7]

The Miskito managed to stay outside of Spanish domination partly because the Spanish found little of interest on the Miskito Coast, and partly no doubt from the Miskito's ability to put up a good fight. Long (1774) asserted that the Miskito "have bravely maintained their independence and keep alive an inveterate abhorrence of them [Spanish], by reciting, at their public councils and meetings, examples of the horrid cruelty practised upon their brethren of the continent [p.317]."

No wonder that a mutual dislike developed between the Miskito and the Spanish. Dampier (1968:16) related that the Miskito hated the Spanish "mortally." On the other hand, according to Dampier, the Miskito were in general "very civil and kind to the *English*, of whom they receive a great deal of Respect ... [they] acknowledge the King of England for their Sovereign. They learn our Language, and take the Governour of *Jamaica* to be one of the greatest Princes in the World [p.17]."

The Miskito in alliance with English buccaneers made frequent raids up the Río Coco, Río Grande, and Río San Juan against Spanish settlements in the interior of Nicaragua in the seventeenth and eighteenth centuries. These forays forced the Spanish to abandon many frontier settlements and retreat to the west as the towns of Nueva Segovia, Jinotega, Telpaneca, Matagalpa, Sebaco, and Muy Muy were repeatedly destroyed (Conzemius 1932:86–87; Denevan 1961:290–291). For example, the town of Nueva Segovia, then located at what is now Ciudad Antigua, was attacked in 1654, 1704, 1709, 1711, and 1743 by the English and Miskito.[8] In 1789 Nueva Segovia was moved to a new site (Ocotal) further away from the frontier (Denevan 1961:291). These were but a few of the attacks made along the entire Spanish frontier.

Backed by their English allies, and acting under the figurehead of a Miskito King, the suddenly powerful Miskito continued to expand their domain away from their shoreline home, upriver into country inhabited formerly by Sumu Indians. There they had to adapt to a differ-

---

7. Translation by the author.
8. Information obtained from Catholic Church records at Ciudad Antigua, Nueva Segovia, Nicaragua.

ent environmental and ecological situation in terms of subsistence. The Miskito had little or no experience with some of the biota of the interior and many of these were referred to by their Sumu names as were the village sites which the Miskito took over.

Attracted to distant resources for market sale and to the rewards of warfare against various groups, the growing Miskito population at Sandy Bay and Cape Gracias a Dios, and other nearby settlements, began to locate new villages along the coastline, lagoons, and up some of the rivers. Since the men were often away from the villages for long periods of time, either on raids with the English against the Spanish, warring against other tribes, or journeying to the south after hawksbill shell, much of the subsistence base was left in the hands of the remaining men and women. During the men's absences, the women and boys would care for the plantations and make do for food as best they could. Bell (1899), who lived on the Miskito Coast from about 1843 to 1859, recalled that the Miskito women

> lead a sort of picnic life while their men are absent. They stray away to visit their neighbours at the mouths of the adjacent rivers, camp out in the bush gathering oil seeds, wander for days among the mangroves catching blue crabs, or go to some distant lagoon to feed on cockles and oysters. Generally they devote a month to camping on the beach, where they keep an immense pot boiling day and night, making salt from sea-water, and they are generally living on the beach when the men are expected to return [pp.85–86].

> They wander over the bright sunny river, or into the dark, lonely creeks, fishing as they go, and with the help of young lads and the dogs killing agoutis, pacas, iguanas, tortoises, and such easily captured game [p.263].

Coastal villages remained attractive settlement sites because of the abundance of fish and turtles, and the opportunity for contact and trade relations with Europeans. However, the Miskito population, still centered on the northeast coast of Nicaragua, was increasing and putting a strain on local food resources. Roberts (1827), writing of conditions around Sandy Bay, indicated that the Miskito had to travel far inland from their villages to a place called "The Hills," in order to obtain agricultural produce which supplied

> the people at Sandy Bay, Cape Gracias a Dios, and other places on the coast, with the greater part of their provisions, such as bananas, plantains, etc. Being too distant from the coast to combine the advantages of agriculture, with those of fishing and trading, no strangers have yet settled on this high ground [p.142].

This, coupled with a poor agricultural base, and frequent long absences of men, created an unstable subsistence situation. The Cape

Gracias–Sandy Bay area, not able to support the suddenly opened economic system and the increased population, became a center for out-migration to other more favorable places,[9] including newly conquered territory.

During the 300 years since the first recorded estimate of Miskito population, their numbers have increased and spread dramatically (Table 1). Early population figures for the Miskito are confusing because many of the estimates are not comparable. The overall pattern, however, from the late seventeenth century to the present day, is one of increase with some reductions due possibly to warfare and disease.

Young (1847) claimed that the Miskito once suffered a large loss of population due to smallpox: "Some years back the smallpox carried off great numbers but latterly they have escaped this visitation. It appears that this country was once thickly populated, and that it was to this awful malady, and not to internal wars, the reduction in their numbers is attributable [p.73]."

TABLE 1

Miskito Population Estimates 1671–1969

Year	Population	Source	
1671–1672	1600–1700 total (Nicaragua)	Exquemelin	(1856:167)
1681	no more than 100 men at Cape Gracias a Dios	Dampier	(1968:15)
1699	1000 total (400 in Sandy Bay)	M.W. Conzemius	(1732) (1932:13)
1725	2000 men	Lade	(1744)
1806	1500–2000 men	Henderson	(1811:225)
1899	5000–7500 total	Bell	(1862:250)
1932	15,000 (Nicaragua and Honduras)	Conzemius	(1932:13)
1969	35,000 (Nicaragua only)	Nietschmann	(1969)

TRADE AND TURTLES

The first prolonged contact between the Miskito and Europeans began in about 1634 when English colonists from Providence Island set up a trading station at Cabo Gracias a Dios. Just south of the Cape was a good harbor where ships could anchor and there the trading post was built, surrounded by a palisade. "The trade was undertaken on a very systematic fashion, small parties being sent out to all Indian villages within reach. The goods issued out of the store to each of the parties

---

9. For example, Little Sandy Bay, just north of Río Grande Bar, was established by Sandy Bay people, as was Tasbapauni.

were entered in a register [Newton 1914:165]." Private trade between individuals and the Indians was discouraged by the company and "they even attempted to put a stop to the trade in parrots and monkeys by charging the sailors ten shillings apiece for their freight [Newton 1914:166]." The traders were cautioned against aggression and to carry on business "by way of peaceful commerce [Newton 1914:165]." The trade station lasted until 1641 when the Spanish captured Providence Island.

Thus began trade relations with the Miskito which were later to be expanded by contact with European buccaneers, English settlers and traders, Cayman Island turtlemen, and, starting in the late nineteenth century, American lumber and banana companies. Contact and trade were to have significant influences on the Miskito's culture, their subsistence, and on their environment.

From the early seventeenth century to the early twentieth century, trade relations between the Miskito and foreigners, principally the English, were based on extractive resources such as tropical woods, especially dyewoods; "skins of all beasts that have any fur or seem vendible," particularly jaguar and deer skins; hawksbill shell; dried green turtle meat; sarsaparilla; gum of pine trees and other gums; chinaroot; anatto (*Bixa orellana*); silkgrass (*Aechmea magdalenae*); rubber; indigo; cacao; and canoes and paddles (Long 1774:319; Roberts 1827:109; Newton 1914:148–149, 166). In exchange, the Miskito would receive lines, cotton clothes, machetes, knives, axes, saws, nails, fish hooks, cooking pots glass beads, rum, gunpowder, muskets, and fowling pieces (Roberts 1827:300–301). The acquired need for foreign trade goods prompted the Miskito to range far and long to find the sought after forest and sea resources. This increased their knowledge of the countryside and helped them choose favorable new settlement locations for their expanding population. Oftentimes the Miskito acted as middlemen in order to obtain trade items from Sumu peoples (Bell 1899:266). Because of this and the control they held over the coast, Miskito soon became the lingua franca of the Sumu subtribes.

Trade relations and market opportunities motivated the Miskito to extend their hunting, fishing, and gathering pursuits beyond subsistence needs to exploitative enterprises. Previously little used or ignored animals such as hawksbill, crocodile, caiman, and jaguar came to be valuable animals. Also, more pressure was put on meat animals such as manatee[10] and green turtles in order to supply the buccaneers, traders, and foreign residents on the Miskito Coast.

---

10. Dried manatee meat was exchanged by the Miskito Indians to British traders for export to Jamaica (Roberts 1827:96; Sloane 1709, II:329). Quoted in Parsons 1956:63, footnote 8.

Essential to an understanding of coastal Miskito subsistence and European trade relations is an awareness of the importance of sea turtles, particularly green turtles, in sustenance and commerce in the Caribbean and along the Central American coast. Once abundant and widespread, turtles provided food for coastal indigenous peoples, sailors, planters, and slaves. As Carr (1956) observed:

> More than any other dietary factor, the green turtle supported the opening up of the Caribbean.
> It had all the qualities it needed for a role in history. It was big, abundant, available, savory, sustaining, and remarkably tenacious of life. It was almost unique in being a marine herbivore—an air-breathing vertebrate which grazed submarine beds of seed plants as the bison grazed the plains and which, like them, congregated in tremendous bands. It was easy to catch with simple equipment because its pastures lay under clear shallow water [p.240].

The green turtle was a major dietary staple of the Miskito Indians and much of their subsistence system, settlement patterns, and scheduling of activities were geared to the spatial and temporal occurrence of turtles. The Miskito were probably the best turtlemen in the Caribbean. Their ability and skill in turtling and sea travel were superb and they quickly attracted the attention of the English who may well have learned the art of turtling from the Miskito (Parsons 1962:30). This was to have a profound impact on the sea turtles of the Caribbean for it was "the British who organized the intensive exploitation of the Caribbean turtle fishery [Parsons 1956:33]." For more than 200 years, relations between the coastal Miskito and the English were often over turtles.

The English focused their exploitation of green turtles on the turtle grounds off the Cayman Islands and the Miskito Coast. The first reference to English interest in commercial turtling activities was in 1635, when the London directors of the Providence Company at Cabo Gracias a Dios expressed concern "that the turtle should fail at the Mosquitos [Parsons 1956:33]." By 1671, there were Englishmen living on the Miskito Cays, the focal point of Western Caribbean turtle fishing (Sloane 1709, I:xvii, xxvii). Turtling activities increased as the English settled more and more of the Caribbean and demands rose for meat to feed sugar plantation workers. Ships from Jamaica annually visited the Nicaraguan Coast by 1722 to catch and buy green turtles, along with hawksbill shell, from the Miskito Indians (Parsons 1956:33; Fernández 1881–1907, IX:155). The Cayman Islands became the center of turtle exploitation in the Caribbean in the late seventeenth and eighteenth centuries and the English settlers on the Caymans soon excelled as turtlemen. For some 200 years the Cayman area supplied green turtles to ships of all nations and to inhabitants in cities and plantations in the Caribbean (Parsons 1962:27–29).

Exploitation was so intense that the nesting beaches and the turtle populations were decimated and the Cayman turtlemen began to turn to other turtle areas off Cuba, the Gulf of Honduras, and the Miskito Coast. They may have started turtle fishing off the Miskito Cays as early as 1837 (Doran 1953:165) and by 1842 were making regular visits according to Thomas Young (1847) who wrote that Cabo Gracias a Dios

> is often visited by small schooners, from the Grand Cayman's island, near Jamaica, to fish for turtle near the Mosquito Kays, about forty or fifty miles from the Cape, and which seldom return without a rich harvest. They supply the Belize and Jamaica markets with the finest green turtle, and often obtain in a season, several backs of hawk's-bill turtleshell; as the Mosquito Kays are very much the resort of that species, as well as of the green turtle. Numbers of the Mosquito Indians sail in their doreys from the Cape, Sandy Bay, Duckwarra, Warner Sound, etc., to fish there. When they find the turtle on the beach, they turn them upon their backs, or spear them as they float on the top of the water [p.17].

Bell (1899) observed that by the end of the nineteenth century, "almost all of the turtles used in Europe come from the Mosquito Shore [p. 41]."

The Miskito used to make long-distance voyages in seagoing dugout canoes to Costa Rica for hawksbill shell; there being few hawksbill off the Miskito Cays, but large numbers to the south, especially off the mouth of the Río San Juan at some submerged rocks called the "Greytown Banks." Bell (1899) noted that "every year, about February, the men used to depart for the hawksbill turtle fishery, returning in May. . . . These turtlemen always went to the southward for their fishing, and were therefore called 'southward men' [pp. 19–20]." The Miskito usually returned to Nicaragua by May when green turtle could be easily taken during the calm weather and calm sea.

During the Miskito's raiding expeditions to the south they often stopped at the Tortuguero, Costa Rica green turtle nesting beach between July and September, in order to procure fat from boiled down turtles caught nesting and turtle eggs. The eggs were dried in the sun to preserve them, "and in this way many thousands of turtles are annually destroyed or prevented from coming to maturity [Roberts 1827:93]." Large amounts of dried green turtle meat, turtles, and hawksbill shell were sent from the coast.[11]

---

11. For example, hawksbill shell was exported from the Coast at the rate of 6,000–10,000 pounds every year during the early nineteenth century (Hodgson 1822 ed.:17). One hawksbill turtle averaged 4 pounds of shell. By the mid-nineteenth century the shell was worth $6.00 per pound (Bell 1899:276).

Scattered green turtle nesting along the mainland was overexploited and nesting activities ceased. Inhabitants of Sandy Bay still speak today of a nearby mainland nesting beach for green turtles that is long gone because of overexploitation. Parsons (1962) mentioned an 1884 proclamation from Bluefields prohibiting "the thoughtless and improvident practice of destroying the nests of turtles for the purpose of carrying off their eggs [p. 32]."

The underwater marine pastures off Tasbapauni were an important turtling area for the Miskito and Cayman Islanders. Bell (1899) remarked that in May, the height of the turtle season,

> frequently a hundred canoes assembled on the Man-O'-War and King Keys, where the Indians encamp under the [sea] grape-trees surrounded by an enclosure of live turtle.... Along the beach, between the grape-trees, lines are stretched, which are covered with turtle meat drying in the sun. The whole island is strewn with fragments of meat, and the smell of the green fat fills the air and is smelt miles to leeward [p.274].

Cayman Islanders continued to net and to buy turtles from the Miskito Indians and to supply foreign markets until the 1960s when they were unable to renew their fishing privileges with the Nicaraguan government. The Cayman turtlemen introduced turtle nets to the Miskito who had, up to that time, taken turtle only by harpoon. One ancient informant in Tasbapauni claims that "de older heads dem used to strike de turtle, no bodder wid de nets til Caymansman bring in dat net work. Learned de Indians how to make and set."[12]

In the mid-nineteenth century two events took place which had a great impact on the Miskito: the coming of the Moravian Church in 1849 and the Treaty of Managua in 1860. A few years after its establishment in Bluefields the Moravian Church began to send representatives into Miskito villages to christianize the Indians. From 1855 on, church work among the Miskito spread rapidly. By 1900 almost all of the larger coastal Miskito villages had a resident lay pastor and received periodic visits from Moravian missionaries. Helms (1971:182–216) discusses the importance and significance of the Moravian Church in the life of the Miskito. Clearly the Church has reshaped much of the outlook and interrelations among the Miskito. Moravian missionaries also introduced some new crops and agricultural techniques. Many fruit trees such as breadfruit, rose apple, and star apple were spread from village to village

---

12. Several excerpts from personal conversations with Miskito informants will be included in this study. The comments were transcribed from tape recordings made in the field. If the individual was speaking to us in Creole the transcription is direct; if the conversation was in Miskito, the comment was translated into Creole, trying to keep the meaning and flavor as near as possible to the original.

by the missionaries. Changes in marriage rules, world view, and division of labor occurred as a result of the Church.

England was forced by pressure from the United States—who feared England's influence in the increasingly strategic proposed canal route up the Río San Juan—to sign treaties in 1859 and 1860 with Honduras and Nicaragua granting them control over the Miskito Coast. Nicaragua created a "reservation" of a portion of the coast which lasted until 1894 when the area was incorporated into Nicaragua as the Department of Zelaya. The Nicaraguan government effectively cut off a good deal of outside English trade, which had been a mainstay of the Miskito economy for more than 250 years, and attempted to bring the coast under national control. Because the Miskito and Spanish-speaking peoples had a history of very unfriendly relations, much friction occurred between the two groups. The Miskito are not yet entirely under effective national control.

As a result of the long contact with buccaneers and traders, the Miskito became one of the first acculturated groups along the Caribbean coast of Central America. They assumed a superior power position over other Indian groups on the coast and were able to dominate all of the eastern tribes from Honduras to as far south as Chiriqui Lagoon in Panama. Their subsistence food system was altered somewhat by the long absences of men from the villages while looking for sea and forest products to be traded. Similarly, aboriginal reciprocal economic exchange patterns were changed through European contact. Additional work effort had to be directed toward acquiring forest and sea products which could be traded for desired goods. This resulted in a reorientation of some labor and exchange patterns in the community and a decrease in the time available for food-getting. Increased exploitation of particular species beyond subsistence needs in local ecosystems had the same ecological impact as if the Miskito's population had suddenly and drastically increased.

COMPANY PERIOD

The establishment of foreign-owned companies on the Miskito Coast intensified economic influences from outside systems on the Miskito: contract wage labor instead of trade exchange; long-term employment opportunities instead of the previous seasonal, short-term economic relationships with foreigners; and a more consistent impact on ecosystems and cultural systems.

Commercial lumbering activities carried out on a limited basis in the eighteenth century with the cutting of mahogany and other valuable woods, was expanded in the nineteenth century. Several mahogany companies began operations along the coast, hiring crews of Miskito Indians to locate, cut down, and bring out the trees from the forest.[13] By the end of the nineteenth century, wild rubber collection and mahogany operations were the two most important commercial activities in the Tasbapauni area, with several banana companies just beginning plantings along the river floodplains (Fig. 5).

The banana era was the most important economic period on the Coast for the Miskito, lasting from about 1890 to the 1930s, but Panama disease and Sigatoka disease caused a decline in the plantings and finally abandonment. Independent buyers continued to purchase bananas from local growers along the Río Coco, Río Grande, and Río San Juan up until the 1950s. Also starting in the late nineteenth century, gold mines in the Siuna–Bonanza area began to employ Miskito workers.

Lumbering and gold mining have kept going with peaks and depressions to the present day. The northeastern pine savannas began to be exploited in the late nineteenth century; large-scale export lumbering started in 1921 with the opening of Bragman's Bluff Lumber Co. (Parsons 1955a:55–56). Extensive activities ceased after 1960 when Nicaragua conceded its claim to a large portion of land north of the Río Coco to Honduras which had the largest remaining pine stands.

The various wage employment opportunities with the many foreign companies engaged in extractive exploitation of natural resources and plantation agriculture attracted Miskito men away from their villages (Figs. 6 and 7). During some periods, jobs were abundant and companies competed for laborers; while at other times, company work was scarce and jobs were few. The boom-and-bust economic period that characterized the late nineteenth century has continued up to today, with the Coast now being in a 20-year-long "bust" period.

Miskito men often went away for a year, taking wage contract jobs with lumber or banana companies, and sending money and commissary goods to their families in the village, sometimes returning only at Christmas. Anyone who went away from the village to work for a long period was known as a *mani uplika* or "one year man."

For the Miskito the company period brought increased culture contact, introduction to a money-based economy, increased periods of absence of males, exposure to relatively inexpensive foreign commissary goods, and an increased knowledge of the total environment as a result of their experiences as rubber tappers, lumber scouts, and trail cutters in the selva.

13. For example, the George D. Emery Company of Boston had almost a monopoly in mahogany lumbering from 1894 to 1902, exporting 1000 logs per month from Río Grande to Boston (Parsons 1955a:55).

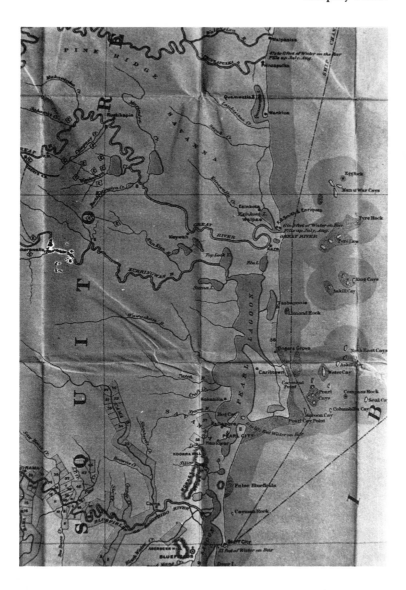

Fig. 5. An 1894 map of Tasbapauni area and zones of exploitation. The diamond-shaped symbols with an "E" represent the location of George Emory's mahogany camps. The numbered areas along the Río Escondido indicate early land division for banana plantations [from "Map of the Mosquito Coast," compiled by H. G. Higley, 1894].

Fig. 6. Miskito men loading bananas on the Wawasang River (near Tasbapauni) in about 1915 [from "Casa Aleman" Moravian postcards].

Fig. 7. Cutting pine trees on the northeastern savannas in about 1915 [from "Casa Aleman" Moravian postcards].

The periodic absence of men working for lumber and banana companies affected subsistence activities in the villages. The women did most of the work in agriculture, sometimes hiring available men to do the hard work of felling and clearing the forest for a swidden site. Old men did the hunting and turtling. Cash crops, such as rice, bananas, and plantains were cultivated by villagers to supply company food needs.

The many years of contact and monetary exchange with foreign traders and companies initiated an acceptance of a cash-based exchange economy in Miskito society which operated alongside, but often in competition with, the subsistence economic system. Where the subsistence economy was based on social responsibilities, the cash system had as its basis, unit value exchange of resources and labor. Men sent money and goods back to the villages to substitute for their previous contribution to subsistence. Money was used by the remaining villages to keep the subsistence system going through purchase and hire. Labor relationships began to include wage labor in addition to the traditional communal exchange of labor (*pana pana*; "hand to hand") and cash sale of produce in addition to traditional generalized reciprocity.

Miskito cultural and social patterns had to be adjusted in order to adapt to a money economy and to the commercial and economic opportunities from traders and companies. Helms (1969b:328–29, 340) terms the Miskito a "purchase society" as a result of their adaptation. Existing outside of the sphere of state political control, but within the economic frontier, the Miskito were engaged with a wider society solely by trade and wage labor. In order to facilitate participation in the money economy, cultural adaptations took various forms, such as an alteration of traditional agricultural labor exchange to permit wage labor arrangements, and extra effort and cooperation to obtain forest and sea products, and agricultural produce to exchange for money or goods.

The presence of foreigners and foreign companies had a devastating effect on the environment. Portions of the tropical forest have been modified by logging operations and most of the mahogany and cedar are gone. The once extensive stands of pine on the northeastern savannas have been cut. Animal populations long hunted for market sale of meat, skins, and shell have been severely reduced. Heavy exploitation of green turtle nesting beaches in the Caribbean, especially at Tortuguero, along with high catches of turtles by visiting Cayman Islanders, and the large-scale sale of turtle meat to supply banana and lumber companies, have greatly reduced the size of the green turtle population off the Miskito Coast. Informants in Tasbapauni can recall in the early twentieth century when 50 turtles could be easily taken in 1 day. Now the catch is much smaller; the turtlemen have to go farther, stay out longer, and then often come back empty-handed.

The exchange of resources from the Miskito's ecosystem to extraneous systems created ecological alterations while the participation in trade and cash exchanges between the Miskito and foreigners created cultural and social alterations. The economic mainstay of outside inputs of jobs, currency, and inexpensive goods has been removed. The company commissaries and supply boats are gone, money is scarce, and store-bought items are expensive. With few outside work opportunities the Miskito have become increasingly dependent once again on their own subsistence efforts and on their own local environments for salable resources, but this time under greatly changed social and ecological conditions. The golden boom period is past and the Miskito are left with only their subsistence system together with an overburden of desires for luxury and foreign goods as a result of contact. This has created an "ethic of poverty" where "foreign objects and luxury foods obtainable through participation in the market economy are no longer available. Hence life is not completely satisfying. A feeling of want and of isolation has become dominant [Helms 1967:235]."

The Miskito today are highly acculturated in many outward appearances (Figs. 8 and 9): They participate in a market economy; they often speak Creole English, a few speak Spanish; most are members of the Moravian, or to a lesser degree, the Anglican (Episcopal) or Catholic church; some have a grammar school education; and they are very much aware of the "outside world." However, throughout the early contact period of trade with English and other foreign people, and through the turbulent boom-and-bust cycles of the late nineteenth to mid-twentieth centuries, the Miskito have maintained many of their traditional cultural patterns, including their language,[14] and most of their subsistence system. After a very thorough review of the economic history of the Miskito, Helms (1967:28–29; 1969b) is of the opinion that the continued existence of an identifiable Miskito culture is due to the fact that they have functioned in a boom-and-bust economic frontier, rather than in an agricultural frontier zone, which would have meant competition for the land base, destruction of the forests, and disruption of Indian cultural patterns as it has elsewhere in Latin America. Under this veneer of acculturation, then, much of the Miskito culture and society retains traditional attitudes, behavior, and technology, especially toward the procurement, preparation, and consumption of food.

---

14. Although there are five dialectical variations among the Miskito, differences are slight, and the Miskito consider themselves as "one people." Of the five recognized Miskito dialects, four occur in Nicaragua: *Wangki bila, Tawira bila, Kabo bila,* and *Baldam bila.* The Miskito in eastern Honduras speak *Mam bila* (Conzemius 1929a:59–60).

Fig. 8. Miskito family in about 1910. The photograph was possibly taken in Tasbapauni [from Moravian postcard series].

Fig. 9. Members of a Miskito family in 1969, Twappi.

45

CHAPTER IV

POPULATION AND SOCIETY

The location, size, and age–sex composition of Miskito population, their state of health and medical problems, and the organization of the social and economic system are factors to be discussed in this chapter which will be relevant to later analysis of subsistence production and distribution and human pressure on ecosystem resources.

POPULATION

Population Size and Location

The Miskito population in Nicaragua was estimated to be approximately 35,000 in 1969 (Nietschmann 1969). The coastal population numbered 10,000–11,000, extending from Old Cape in the north to Haulover at the southern end of Pearl Lagoon. The Río Coco Miskito had a population of 14,000–15,000, distributed along the river for more than 300 miles. Most of the remainder of the Miskito were located along the Wawa, Kukalaya, Prinsapolka, and Río Grande rivers, and in the Siuna–Bonanza gold mining district (see Fig. 3, p. 16). The Miskito also inhabit the coastal lowlands as far north as Cabo Camarón in Honduras (Helbig 1965). Their overall distribution then, from Cabo Camarón to Pearl Lagoon, a distance of over 400 miles, and more than 300 miles up the Río Coco, makes the Miskito the second most widely distributed Indians, after the Maya, in Central America.

Most of the Miskito now live away from the coast, their former hearth area, and occupy locales formerly inhabited by Sumu Indians. Approximately 30% of the Miskito population reside on or near the coast. The rest are riverine peoples who have had to adapt to new surroundings, environmental conditions, and food resources.

The Miskito population is expanding and new settlements are being established in many places, especially along the Río Coco. In addition, the recent influx of displaced Miskito from Honduras, who moved to Nicaragua after the boundary settlement in 1960, has created new villages along the Río Coco.

Most of the Miskito villages have a population of 200–400. A few riverine villages such as Raiti, Asang, and Saupuka on the Río Coco have over 700; while some coastal villages such as Ninayeri (one of the 12 Sandy Bay villages), and Auastara have a population of 400–500. Tasbapauni, one of the largest Miskito villages, has nearly 1000 inhabitants.

The distribution of the Miskito population and the location of their villages, as shown in Fig. 3, points out the contrasts in the relative population densities between the Río Coco Miskito and the coastal Miskito. Because of the small map scale, not all of the Río Coco villages can be shown; nevertheless, it can be seen that their density is much greater than those along the coast. With the exception of the 12 communities that make up the Sandy Bay settlement, coastal villages are quite spread out, with long stretches of intervening swamps and lagoons. Almost all the coastal villages are situated at or near a river bar, at a haulover, or on a river or lagoon with ready access to the sea. This distribution is important in terms of the degree of intensity to which the available food resources are exploited. One good, but localized hunting area, can be quickly depleted if a number of villages have access to it. The concentrations of coastal Miskito villages around Sandy Bay to Auastara in the north and from Little Sandy Bay to Set Net in the south are a result of, and are heavily influenced by, the nearby occurrence of the two largest turtle grounds.[1] Tasbapauni has a large population probably due to its haulover location and ease of access to land and sea resources.

Almost no reliable population data are available for the Miskito Indians. Readily obtainable government census figures are for large political *departamentos,* and information on smaller areas is very general

---

1. One observer in the early nineteenth century remarked of the Old Cape settlement that "it is at no great distance from the Mosquito Kays, whence can be procured, at all seasons, inexhaustible supplies of the finest green turtle; and, but for this last circumstance, the Cape would be, perhaps, entirely deserted. . . [Roberts 1827:151]."

due to their inaccessibility and because large segments of one village's population are often away planting in distant agricultural sites or away seeking work when the census was taken.

The population totals for the coastal Miskito villages shown in Fig. 10 were compiled from personal reconnaissances along the coast during which head and house counts were made.[2] Information was also obtained from missionaries, field personnel of the malaria control program, and the government cartography department. These figures are to be taken as estimates only, not as precise population data. I do believe, however, that they are more reliable than any other data currently available.

The age–sex population structure for the study village of Tasbapauni is shown in Fig. 11. These population data were collected during the beginning of the rainy season in July 1969, when most of the villagers were at home. Younger individuals knew their ages but many older people had only an approximate idea, relating their age to when a hurricane passed, or to when a foreign company started or closed. The age–sex composition points out that the population is young; almost 60% is under 21 years of age. The low number of adults, especially males, in the 20- to 40-year tiers of the population pyramid is the result of their absence in search of work opportunities away from the village. Even though wage labor jobs are extremely scarce for the Miskito, their need for sources of cash is forcing some individuals to "look work" on shrimp boats, in the Siuna–Bonanza gold mines, or in Bluefields or Puerto Cabezas. The imbalance of individuals in these age groups places additional pressure on others to produce and distribute food in the village.

The amount and degree of subsistence production is roughly correlated to age categories. It must be understood that for the Miskito, a general subsistence rule is that one should work as long and as hard as necessary to supply food for family and extended family responsibilities. When an individual becomes old, he or she continues to work doing heavy agricultural labor, or in the case of men, hunting and turtle fishing, well into their 60s and even 70s. One is a productive member of Miskito society according to their ability and social responsibilities. The distribution of age groups according to rough categories of economic

2. A figure of seven persons per house was used to estimate the population in many villages where a quick head count would have been thought of as extremely rude. A visitor is expected to sit at each house awhile and have a bowl of *wabul* (banana porridge; usual serving is about ½ to ¾ quart) or a water coconut. Quick "hit and run" research is a little abrasive to the Miskito. Given only a few short days in many of the more remote communities, and having other research objectives, I often made only house counts. The average number per house was obtained from samples in other communities where more precise population estimates were made.

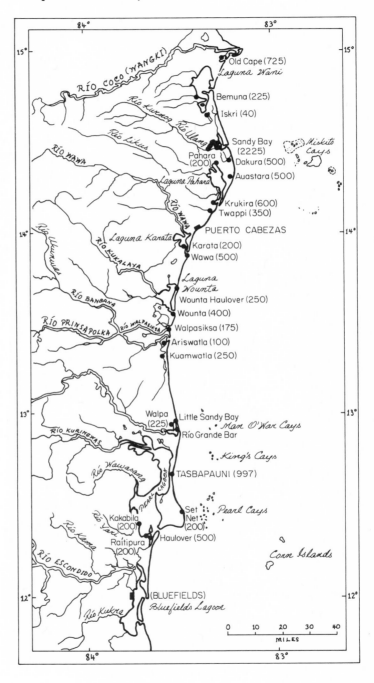

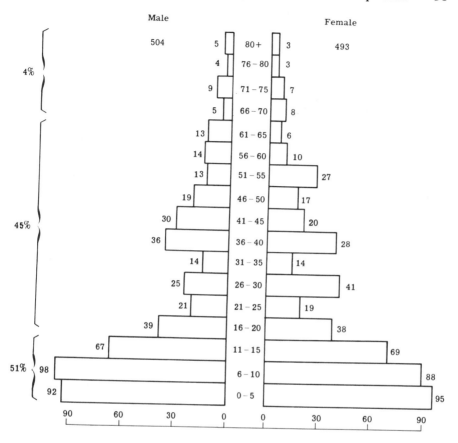

Male   Female

504   5 ☐ 80+ ☐ 3   493

4%   4   76 – 80   3

                        9   71 – 75   7

                        5   66 – 70   8

                       13   61 – 65   6

                       14   56 – 60   10

                       13   51 – 55   27

                       19   46 – 50   17

45%   30   41 – 45   20

                       36   36 – 40   28

                       14   31 – 35   14

                       25   26 – 30   41

                       21   21 – 25   19

                       39   16 – 20   38

                       67   11 – 15   69

51% { 98   6 – 10   88

       92   0 – 5   95

90   60   30   0   0   30   60   90

Fig. 11. Population structure of Tasbapauni, July 1969 (997 total).

production is shown in Fig. 12. Young children, 0–5 years, are generally completely dependent on others for sustenance. Older children 6–15, help supply some food and materials for the community's needs. Together these two groups comprise 51% of the village's population, yet contribute little to subsistence. As one Miskito said of his 13-year-old-son: "All his turtling is on de plate." Between the ages of 16 and 50, which make up 36% of the population, are the most productive years. It is these individuals who are primary producers and distributors

Fig. 10. Coastal Miskito villages and size of population, 1968–1969.

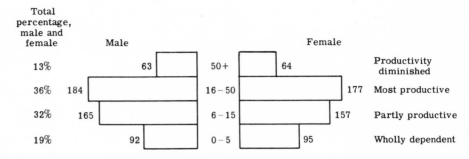

Fig. 12. Distribution of population by economic age group, Tasbapauni, July 1969.

of agricultural foodstuffs and meat from hunting and turtle fishing. After 50 years of age, even though work continues, productivity is diminished.

The wide base of the younger age groups in the Tasbapauni population pyramid indicates a rapidly growing population. Most of the indigenous methods of birth control such as female infanticide, abstinence from intercourse during breast feeding of children until they were 2, 3, or 4, have long been changed as a result of intervention from churches and the "modern world." The Miskito say that families are larger today than they used to be at the turn of the century.

No population increase figures are available for the Miskito, but judging from the number of births and deaths in Tasbapauni during field research, the number of new houses in villages, and estimates made by informants of previous village populations, the population must be increasing at least at a rate of 3.0% per year, and probably closer to 3.5%.[3] By projecting Tasbapauni's 1969 population of 997 at an estimated 3.3% annual increase, and assuming for the moment that there is no out-migration, the village will have a population of 1383 in 10 years, and 1913 in 20 years; thereby doubling exploitative pressure on the ecosystem for human energy needs.

Body Size

Coastal Miskito are usually much larger in body size than Río Coco Miskito. This is probably the result of two major factors: higher Negro admixture and a generally superior diet which is much higher in protein

3. Estimates of population increase rates for Amerind groups are exceedingly rare. Bennett (1968:58–59) noted an annual increase during the period 1950–1960 of 1.1% for the Cuna (Panama) and 4.1% for the Guaymí (Panama).

intake. Body size data obtained by Pijoán (1946, 1:53) for Río Coco Miskito show that the average height of 19-year-old males was 63.8 inches and for females 63 inches. Pijoán's (1946, 1:53) weight estimates for Río Coco Miskito were 136–154 pounds for men, and 143–160 pounds for women.

Our own observations of size and weight of 10% of the Tasbapauni population are shown in Table 2. The average height of 20- to 40-year-old men was 66.7 inches, with a range of 63.5 to 71.5 inches; average weight was 148 pounds, with a range from 126 to 205 pounds. Women in the same age group had an average height of 60.8 inches, and a range from 56.8 to 66.0 inches; weight averaged 124 pounds, with a low of 111 and a high of 142 pounds.

In general Tasbapauni men are fairly tall in stature, lean and muscular, extremely strong, and have remarkable endurance (Fig. 13). Exactly how much their larger stature and weights are expressions of different diets, genetic background, or parasite infestations, compared to their Río Coco counterparts, remains to be answered.

Fig. 13. Miskito turtlemen, Tasbapauni, 1969.

TABLE 3

Body Measurements Based on 10% Sample of Population, Tashapauni, May, 1971[a]

		0–3 years ♂	0–3 years ♀	4–12 years ♂	4–12 years ♀	13–19 years ♂	13–19 years ♀	20–40 years ♂	20–40 years ♀	40–60 years ♂	40–60 years ♀	60 + years ♂	60 + years ♀
Weight (lb.)	Avg.	17	24	41.6	44.9	101	102	148	124	133	127	149	133
	Range	12–25	22–25	26–71	21–83	60–150	65–134	126–205	111–142	118–149	108–151	106–191	101–193
Height (inches)	Avg.	26.7	33.2	44.1	44.9	60.7	60.9	66.7	60.8	64.8	62.7	65.7	61.3
	Range	23.0–30.5	32.8–33.5	35.0–56.2	32.0–57.8	52.5–69.8	52.2–69.2	63.5–71.5	56.8–66.0	62.0–67.5	59.0–68.0	62.0–69.2	57.2–63.2
Chest (inches)	Avg.	19.3	20.2	23.6	23.4	31.1	32.0	37.6	36.6	36.5	36.2	38.4	39.0
	Range	18.8–20.8	20.0–20.5	20.5–27.0	19.5–29.5	26.5–37.0	27.0–36.5	35.0–43.0	33.5–39.2	35.5–37.5	33.8–39.5	35.5–43.5	33.0–43.8
Head (inches)	Avg.	17.5	19.1	20.1	19.8	21.3	21.7	22.7	21.3	21.9	21.7	21.9	22.2
	Range	16.5–18.5	18.8–19.2	19.0–21.2	18.0–22.2	20.2–23.0	20.2–22.8	21.5–24.0	20.5–21.8	21.2–23.0	21.0–22.5	21.5–22.5	20.8–23.0
Arm (inches)	Avg.	5.7	5.9	6.4	6.8	9.0	9.2	11.5	11.0	10.8	10.8	11.2	11.2
	Range	4.8–6.2	5.2–6.5	5.2–8.2	5.0–9.0	7.0–10.8	7.0–11.0	10.2–14.5	10.0–12.0	9.5–12.8	10.5–11.5	8.8–12.5	9.0–15.8
Sex ratio		1.33		1.23		0.83		0.64		0.75		1.00	
N		4	3	27	22	5	6	7	11	6	8	5	5

[a]This table was compiled with the help of Mr. Brian Weiss.

HEALTH

Plenty time doctor no help but bush medicine help plenty.

—An Old Miskito

An evaluation of the health status of Tasbapauni villagers was difficult to do because of the lack of medical diagnosis. Most health complaints, however, centered on diarrhea, fevers and body aches, "fresh colds," infected cuts, and occasional dizziness. The Miskito point to the liver, kidneys, heart, and stomach as especially troublesome body organs, and the cause of many of their health problems.

Available medical help is in Bluefields, 70 miles south of Tasbapauni. The long boat trip, expenses for medicine and doctor's services, and difficulty in finding a place to stay in Bluefields, discourage most people from going. Instead most Miskito utilize plant medicines.

The Miskito employ a wide variety of bush medicines made from plant roots, barks, berries, leaves, and ashes. Medicinal plant preparations are usually taken in the form of hot or cold infusions or poultices. Many specific plants are recognized for use as purgatives, antiseptics, and vermicides. Many are used for diarrhea, snake bites, and malaria (Heath, n.d.).

Worm infestation (especially *Ascaris, Trichuris,* and *Enterobius*) and bacillary and amebic dysentery cause most of the diarrhea and stomach disorders, and generally fatigue the body.[4] Malaria is still a problem along the coast, albeit a much-reduced problem from former days, and appears to be making a comeback. Malarial personnel from the Servicio Nacional de Eradicación de Malaria (SNEM) used to visit villages every 6 months to spray house interiors with DDT.[5] In addition, other SNEM officials made more frequent visits, passing out malaria suppressive pills (Chloroquine, Camoquin, and Primaquin) and taking blood samples from individuals complaining of fevers and body aches. The occurrence of malaria appears to be increasing, due perhaps to the decline in SNEM personnel because of economic cutbacks and to more resistant strains of *Plasmodium.* Malaria fevers and symptoms are commonplace, sapping the strength and energy of many villagers. Individuals living directly on the beach front usually escape malaria because the prevailing sea breeze keeps *Anopheles* mosquitos back from the houses. Families living to the leeward of the beach, frequently on the edge of a swamp or

4. Personal communication, Dr. Edwin Wallace, Gray Memorial Hospital, Puerto Cabezas, Nicaragua, 1969.

5. Sprayings of DDT have killed off practically all house-cats in the villages, which are extremely susceptible to the insecticide. With their major predator gone, rats have multiplied and are causing problems with stored food.

dense bush, are often not so fortunate. Malaria is also a major medical problem on the Río Coco. Pijoán (1946, 2:174) reported that in San Carlos, a Miskito village on the mid-Río Coco, the older people occasionally sang a chant:

> *The fever returns and I shall die.*
> *I drink medicine put it on my body.*
> *The fever will kill me and I shall die.*
> *The medicine does not help me.*
> *I have been good to my family.*
> *I have had the fevers.*
> *This time the fever will take me.*

Other major diseases and medical problems for the Miskito include various skin diseases, occasional cases of whooping cough, pneumonia, tuberculosis, malnutrition, especially on the Río Coco, and infected cuts. A recent survey in four Río Coco villages showed that one out of every three adults over 50 years of age had cases of active tuberculosis.[6]

Children are sick more often than adults. A great deal of time and money is spent trying to cure children's illnesses. If bush medicine fails, mothers will try to take the sick child to Bluefields or to Pearl Lagoon for medical help.

The frequent occurrence of fevers, pains, and other ailments, creates debilitating disruptions in an individual's labor pattern in agricultural and meat-getting activities for subsistence, and in the exploitation of forest and sea resources for market sale. Related kin or neighbors will give food to a family if the head of the household is sick.

People in Tasbapauni and in other Miskito villages have widespread dental problems. Whether or not this is related to an inadequate diet will be explored later, but here it is important to note that toothaches and loss of teeth are common, and often cause not only discomfort but pain severe enough to preclude any subsistence activities. The most frequent remedy is to use one of several leaves to dull the pain; later it will be necessary to have the tooth removed or the nerve killed. The latter is the most common method. A nail or piece of metal rod is filed to a sharp point and placed in wood coals until it is red hot. Then the point is placed in the open cavity in the tooth with the aid of a mirror or by someone else. This is said to kill the tooth. An individual who had used this method a number of times told me: "nail hole rotten' de tooth. Quick time come out, piece by piece." Judging from the large number of missing teeth among Tasbapauni people, it must be effective.

---

6. Personal communication, Dr. Peter Haupert and Dr. Ted Rights, Thaeler Memorial Hospital, Bilwaskarma, Nicaragua, 1970.

SOCIAL AND ECONOMIC ORGANIZATION

Remember we have Indian rule here. Indian he don't worry if no have meat for tomorrow. Piece here, piece there—he going to give. Can't eat the money. I have to give to my mother, my aunt, my sister. If you don't like dat, den take another woman.
—a comment made by a Miskito woman to her Creole husband

In traditional Miskito society, social relationships and economic exchange were intertwined in one undifferentiated system. Every social relationship had an economic aspect and every economic exchange had a social context. Miskito interpersonal behavior and relationships were organized and regulated primarily through their kinship system and associated social rules for economic exchange of subsistence foods and labor. Through the many years of culture contact and exposure to foreign cash market opportunities, Miskito social and economic organization has undergone a number of alterations and adjustments in an attempt to fit new patterns to old rules. It is at this interface between the new and the old, between market and subsistence, between customer and kin, that new cultural and social patterns are emerging and new interpretations and use of land and water ecosystems are being created.

Miskito social organization and interpersonal relationships are defined and expressed through the kinship system. Kinship and social networks permeate almost all aspects of Miskito culture and society. The Miskito's emphasis on kinship for social and economic organization was remarked on by Helms (1971) who observed:

> The most obvious feature which strikes the investigator is the use of kinship as an explicit rationalization for much daily activity. Although there is no functioning corporate kin group above the level of the nuclear family, the notion of showing respect for all one's relatives forms the basic code of proper behavior. The constant use of kinship terms and teknonymy in reference and address, together with generosity, particularly in the form of food sharing, are daily reminders of the importance of the kinship ethic. Certain forms of cooperative work, especially house building and slaughtering, are also areas where kinsmen, both consanguinal and affinal, are expected to work together [p. 109].

The Miskito reckon descent through males, terminologically going back no more than three or four generations. Kinship is characterized by recognition of Hawaiian cousin terminology and of bifurcate and collateral relatives. Although village exogamy and matrilocal residence has been the general practice, village endogamy is becoming more prevalent with the increasing Miskito population and desire to marry someone from one's own village (Helms 1967:168–170). The Miskito have accepted the Christian church's marriage traditions, which follow the general rule that one should not marry any relative closer than a third cousin to an individual. This is a general rule; it is not always followed. Polygyny was practiced in some families prior to missionary activity; a man keeping as many wives as he was capable of supporting. This was generally to men of rank or high prestige in the village (Helms 1971:92). Today in Tasbapauni most families are nuclear in organization.

As Helms (1971) noted for the Miskito village of Asang, and true as well for Tasbapauni, "The social structure of Asang is composed of three types of groups: *taya*, the *kiamp*, and the nuclear family, or household [p. 72]." All these groups are kinship based. *Taya* refers to a group which includes all living descendants of an individual's grandparents and their siblings who are considered to be relatives. *Taya* also includes related Miskito who may live in other villages.

Individual Miskito are generally identified as being part of one particular *kiamp*, or surname descendants of an elder male; including only part of the *taya*. Thus, sons and daughters of *Dama* (grandfather) Julius, consider themselves members of the Julius *kiamp*. Oftentimes, members of a particular *kiamp* reside close to their respective family heads. Therefore, members of the Julius *kiamp* may live in different houses in one area of the village, while Carlos *kiamp*, Garth *kiamp*, and Prudo *kiamp* members are situated in other nucleated residence groups. There is no visible boundary demarcation between one *kiamp* and another. Both *taya* and *kiamp* ties are general reference groups for an individual and his or her household serving as "contacts for obtaining food and lodging in other villages when traveling [Helms 1971:75]," and comprise part of the social network in the village through which social and economic obligations are mitigated.

Traditional Miskito culture, kinship patterns, and food distribution are largely precipitated and maintained by Miskito women. Women form a consanguinal core which has kept Miskito cultural patterns intact. The importance of the core of related women in Miskito society for kinship obligations and food sharing is noted by Helms (1970a):

> It is the women who tend to emphasize traditional kinship terms in reference and address, who follow kinship obligations, particularly as represented in the sharing of food . . . when a hunter kills an animal, he requests that his wife send pieces of the meat to his mother and her relatives, and to her own mother's family in preference to those on the man's side of the family [pp. 463–464].

Thus food sharing through a man's mother's side and his wife's relatives reinforces matrilocal residence and social relations among related women in the village. This has been important to Miskito culture history for it maintains strong relations between mothers, daughters, and sisters, providing a stable, unifying point for Miskito culture and society, while allowing men to be free to engage in outside work or in long turtling expeditions. Helms (1970a) suggests that matrilocal residence and the tightly knit group of related women have provided a basis "for the continual local expression of traditional forms of kinship, generosity and general culture patterns [p. 461]."[7]

The Miskito in Tasbapauni denote four major ethnic groupings in the village: Indian, "mixed," Creole, and foreigner (i.e., American missionary, Chinese shopkeeper). Careful differentiation is made between Indian and Creoles in terms of appearance, heritage, culture patterns, and social and economic behavior. In general an Indian is someone who has straight hair, speaks Miskito and only some English, has Indian parents, and follows traditional Miskito customs and social responsibilities. On the other hand, a Creole is someone with pronounced Negroid features, usually speaks English and possibly some Miskito, has Creole parents, and does not conform too rigidly to traditional cultural patterns although social pressures tend to force conformity of behavior, especially if an outside Creole male marries a Tasbapauni Miskito woman. A person referred to as "mixed" usually has curly hair and one Creole parent or grandparent; they are considered culturally as Indians.

In many respects to the Miskito, Creoles personify the outside world with its different economic systems and social responses. The Indians consider the Creoles to be stingy, abrasive and mean, who sell rather than give, who hire people for agricultural work rather than exchange labor communally. If an Indian or a mixed does not honor traditional rules and expected behavior patterns, he or she is thought to have the "Creole way" in them. According to one Miskito informant: "Indian like the monkey. If one fall off a limb, can't get back into the group. Figure have Creole rule into him already; won't let back in."

The majority of the people in Tasbapauni are Indian or mixed; with a minority of Creole families (whether designated culturally or physically). Creoles who reside in Tasbapauni usually adopt Miskito cultural patterns or have to substitute cash exchange for socially determined economic and labor exchange. Most Creole families in Tasbapauni live "Uptown," the northern area of the village, while Indians and mixed live in what is called "Middletown" and "Downtown."

Kinship, however, does not guide all social and economic interaction in Tasbapauni. At the instigation of the Nicaraguan government, village

---

7. Helms (1970a) in the same paper has also suggested that matrilocal residence was a postcontact adaptation to allow the Miskito men to participate in outside economic opportunities for long periods of time.

political organization is enacted through a community board of elected members headed by a "Sindico" (similar to a mayor) who is appointed by the *jefe politico* of the Departamento of Zelaya. The board acts as the intermediary between the village and the State, and is responsible for economic and political decisions affecting the community. It also acts as a mediator in personal disputes over land and property in the village. In addition, practically every Miskito, whether adult or child, belongs to one of the three dominant foreign mission churches: Moravian, Anglican, or Catholic which are organized in keeping with church custom, not Miskito society.

Miskito interaction with local environments is mainly concerned with securing sources of energy and materials which support biological, cultural, and economic needs. The transfer of energy and materials from the ecosystem to the Miskito population and then to each family and individual is organized and carried out through two economic systems: a subsistence economy and a cash economy. Operating alongside the traditional subsistence economy is a market-oriented, cash-based economy, a composite product of trade and wage influences from foreign visitors, settlers, and companies.

Swidden agriculture, fishing, hunting, gathering, and the raising of some domestic animals are the basis of the subsistence economy. Production is by the producers and for their consumption. Small purchases are made of flour, sugar, salt, beans, and other items which are an integral part of Miskito diet and thus a necessary part of life. In lieu of working for companies to obtain money and store goods, the Miskito today are turning to their local environments to acquire products to be sold or traded for desired goods. According to a Miskito informant: "Tasbapauni no have payday like other places. Not like Puerto Cabezas, Bluefields, Prinzapolka. No. Payday here go strike a turtle, go down the beach and back a jute of coconuts to get a five córdoba." Miskito inhabiting the Río Coco raise small quantities of rice and beans as cash crops, collect the sap of the tunu (*Poulsenia armata*) and chicle trees, and hunt or trap jaguars and ocelots for their skins. Similarly, the market economy of the coastal Miskito is based on turtles, skins, shrimp, and small amounts of agricultural crops, principally rice and coconuts. Both riverine and coastal groups raise cattle, pigs, and chickens for sale to outside merchants; very seldom are any domesticated animals butchered in the villages.

The subsistence economy is a continuation of aboriginal (precontact) patterns in the organization of production, distribution, and consumption of foods and other needed materials. The economy is characterized by reciprocal exchange patterns and kinship obligations, involving goods, favors, and generosity. Production activities are formed around domestic units for family needs within a society structured by kinship. Reciprocity of goods and services is carried out between kinsmen and fellow villagers to the extent that what is "economic" is not easily distinguishable from the "social." Family pooling, reciprocity, and a Miskito ethic that to be a good kinsman is to be unselfish, generous, and thoughtful toward others, all interrelate to provide a continuous circulation of labor, food, and materials within a village, the rate of which is geared by need and social relationship, not by Western economic cost accounting.

The introduced cash economy, on the other hand, is recognizable as a distinct system functioning largely between individuals and the outside world, and second, between people in the community. Cash exchange is oriented to supply regional and international markets with forest and sea products where exchange is solely in economic terms, not social. Within the village, where the cash economy functions for some labor arrangements, and the purchase of some food and materials, social considerations shade cash transactions. Helms (1967) observed that "whenever market type transactions occur within Asang, as happens frequently, an effort is made to also fit them into the structure of the subsistence economy. Generally speaking, as long as the subsistence system remains intact, the community will tolerate market transactions within the village. In other words, when an animal is slaughtered, for example, as long as kinship obligations are met with gifts of meat, the rest can be sold for cash or its equivalent [pp. 230–231]." The rest is sold on a cash economic basis, but within a social context. In Miskito society, therefore, money does not necessarily equalize access to resources or labor: Who is selling and who is buying is as important as being able to pay the price.

The relationship between subsistence primacy and subsidiary market sales is changing. Through the long history of economic contact between the Miskito and foreigners, the subsistence system was never replaced by a monetary system, yet this seems to be happening today. Environmental and social pressures are being created by the ascendance of monetary priorities over social or ecological priorities. The exchange economy and the subsistence economy are beginning to operate in direct contradiction to each other, posing new social and economic problems within Miskito villages, and new evaluations of the environment and resource use.

Reduction in outside wage labor opportunities, the increasing need for cash, and declining resources are overloading the subsistence system's capacity to regulate human behavior and environmental exploitation within socially and ecologically optimum levels. Satisfaction of both social responsibilities and the need for small amounts of cash cannot always be realized with available ecosystem resources or available human labor.

Either the resources may be too scarce or individual economic wants too great to permit adequate fulfillment of social responsibilities through the subsistence system. When this happens the system is functioning maladaptively because of environmental depletion or because monetary considerations have gained in importance over social considerations.

CHAPTER V

RAIN, WIND, LAND, AND WATER

This was our portion, earth and sea and sky, the reef beyond the land's edge, a world in itself, the pale green of the lagoons—even as I looked a vaguely moving shadow showed the underwater trail of a giant sea turtle returning from an early morning egg-laying. And the beach itself, the pulsing, moving edge of the sea, that belongs neither to the land nor to the sea, but alternately to one and then the other. These were our portion, these and the soil beyond.

—Gilbert Klingel, in THE OCEAN ISLAND (p. 50)

The Miskito Coast is a complex of varied ecosystems with different associations of fauna and flora offering various combinations of food-getting opportunities in terms of species, site, and season. For the coastal Miskito, living as they do on the edge of the sea, between land and water, ecological factors affecting both marine and terrestrial ecosystems also affect the structure, timing, and success of subsistence efforts. In order to understand how the Miskito's subsistence system is structured and how it functions as an adaptive mechanism, the physical components of their habitat will be discussed in this chapter.[1] Emphasis will be on describing the different ecosystems with which the Tasbapauni Miskito interact, especially the ecological factors of weather and climate, dynamics of flooding and currents, and composition of biotic communities. It would be impossible to describe fully all of the parts and interactions within the various systems; therefore, the concern is with those components and relationships that seemed to me to be functionally related for human subsistence.[2]

CLIMATE AND FLOODS

Seasonal, monthly, even daily distribution and amount of rainfall, the direction and intensity of the wind, and the presence or absence of floods or strong coastal currents are important influences on the distribution and availability of wild fauna, and are critical determinants in the scheduling and ecology of the Miskito's subsistence efforts.

For example, day-to-day weather changes may determine whether or not the coastal Miskito dories can be used to go out to strike turtle or to go hunting. The presence or absence of high winds and rainfall can influence where game animals, especially turtle, will be found. The distribution and duration of periods of decreased rainfall are important for the timing of swidden agricultural activities such as clearing, burning, and planting.

Environmental conditions, especially temperature and rainfall, influence to a large extent the types of biotic components occurring within the Miskito ecosystem. Similarly, soil, rainfall, temperature, and humidity conditions set environmental limits on what types of cultivated plants may or may not be grown, and the degree of success in doing so.

The Miskito Coast is well within the warm, wet zone of the lower latitudes known as the humid tropics. The Caribbean coast of Nicaragua is one of the wettest areas in the American tropics.[3]

Delimitations of the humid tropics are usually based on vegetation and climate. The region is characterized by an evergreen tropical rain forest climax vegetation which requires constantly high temperatures,

1. The *structure* of ecosystems, that is, the spatial and temporal arrangement and combination of biota and environmental elements, will be the focus of this chapter. How the parts of the ecosystems (including human) interact together and affect each other, or their *function*, will be described later. See Clarke (1971:18) for more on the ideas of structure and function in human ecosystems.

2. More extensive descriptions of the physical environment of the Miskito Coast are given by Parsons (1955a) and Radley (1960).

3. See West and Augelli (1966:44) for a map of the distribution of climatic areas in Middle America, and Portig (1965) and Vivó Escoto (1964) for discussions of the climate of Central America.

high relative humidity, and high, fairly evenly distributed rainfall totals.[4] Many areas of the supposed always wet, always humid, always hot humid tropics do, in fact, experience pronounced differences in rainfall patterns and evapotranspiration rates while still supporting tropical rain forest (Grigg 1970:194,199). Although there are probably no areas within the humid tropics that do not have any month without precipitation, there may be one or more months when rainfall is less than potential evapotranspiration.

It is difficult to accurately describe the climate of the Miskito Coast, much less the parameters important to the presence or absence of particular species of biota, because weather data are extremely sparse over time and area. Rainfall records have been kept at Bluefields on and off since 1926 and sporadically at Puerto Cabezas since 1927. Fragmentary or short-term records, kept mainly by mining and banana companies, are available for a few other sites. Reliable temperature records exist for only Bluefields and Puerto Cabezas. Wind data, usually not recorded at the station sites, are obtainable from U.S. Naval Hydrographic Office records from observations made at sea. There are almost no reliable records of evaporation or transpiration rates. Soil, moisture, and soil temperature data are generally absent.

Based on the available records and data obtained in the field over a 1-year period, only a general picture of the climate can be drawn, often requiring interpolation of figures for the very large areas separating station sites. Nevertheless, it will be seen from the discussion and from additional comments made later in the study, that the small-scale variations, breaks and changes in weather, are at least as significant in the ecology of the Miskito subsistence system as the broad general climatic patterns.

Precipitation

When rain, no sun. When sun, no rain. Dats de way it is.
—An Old Miskito

The rainy season here is like the winter in Europe. It brings starvation to man and beast.
—Charles Napier Bell (1899:246)

The entire eastern coast of Nicaragua receives at least 100 inches of rainfall a year, and most of the coast receives more. There is a fairly

---

4. There is considerable dispute over the definition and delimitation of the humid tropics, especially in terms of the amount of seasonality and variations of climatic elements and vegetation types they permit (Grigg 1970:194). Discussions of what criteria delimit and identify the humid tropics are presented by Küchler (1961), Garnier (1961), Grigg (1970), and Blumenstock (1958).

uniform decrease in annual totals from south to north along the coast and from the coastline inland (Fig. 14). San Juan del Norte (the single wettest station in Middle America) is drenched with 250 inches, Bluefields records 163 inches, and Puerto Cabezas 129 inches of precipitation a year. Inland, records show generally 110–120 inches for the coastal lowland.

There are great variations in monthly rainfall totals, and all areas experience relatively dry and extremely wet seasons. There are no rainless months for any part of the coast and therefore the "dry seasons" are characterized by much reduced precipitation levels, generally under 5 inches, long periods of hot, rainless days, often 10–15 days or more, and short rainfalls, lasting only a few hours. Records show that for Bluefields the driest month, March, has an average monthly total of 3.04 inches, well above the 2.4-inch minimal limit set by Köppen as the break between the always wet Af and dry season Am climates[5] (Table

TABLE 3
Average Monthly Rainfall Totals (in inches) for Selected Sites, Eastern Nicaragua[a]

	1890–1892 1898 San Juan del Norte (Greytown)	1926–1932 1934–1949 1953–1968 Bluefields	1927–1950 Puerto Cabezas[b]	1964–1966 Silma Sia	1939–1952 Bonanza
Jan.	22.52	10.42	7.92	6.14	5.76
Feb.	13.25	4.45	3.37	3.74	3.69
March	5.75	3.04	2.41	2.32	2.27
April	13.69	3.34	1.97	1.53	2.55
May	18.25	13.44	8.04	9.93	11.03
June	28.88	19.64	18.03	24.47	19.86
July	34.92	28.20	17.42	15.03	16.57
Aug.	23.80	22.77	15.41	13.76	13.27
Sept.	10.50	13.46	15.81	15.24	12.78
Oct.	21.25	14.42	14.73	11.12	11.69
Nov.	30.74	15.05	13.53	11.90	8.62
Dec.	29.02	15.17	11.09	4.62	9.04
Total:	252.57	163.40	129.73	119.80	.117.13

[a]Source, US Weather Bureau records Bluefields and Puerto Cabezas; FAO records, Puerto Cabezas; Davis 1902; Radley 1960:18; Taylor 1959:57.

[b]More current precipitation records may be available from the National Weather Records Center, Asheville, North Carolina on a "cost search" basis.

5. Two types of climatic regimes characterize the coast: Köppen's Tropical Rain Forest (Af) and Monsoon Rain Forest (Am). The significant difference is the presence of a dry season in the Am climate. Both climates are capable of supporting a tropical rain forest plant cover. The transition between Af and Am climates occurs somewhere near Puerto Cabezas with the Am climate extending north, west, and southwest, while the Af climate occupies a smaller wedge-shaped area to the south gradually widening toward San Juan del Norte.

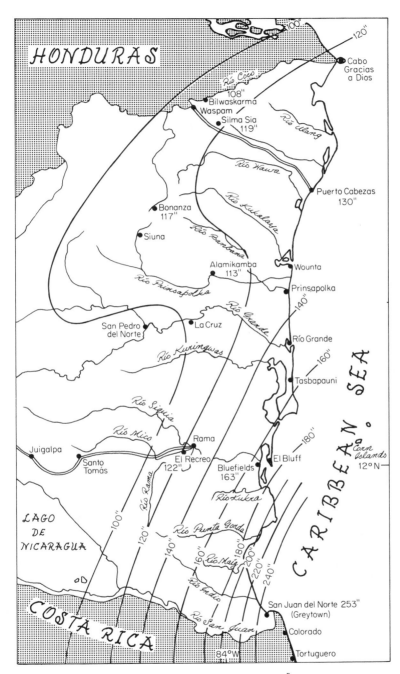

Fig. 14. Distribution of average annual rainfall (in inches) [from U.S. Weather Bureau records in Bluefields and Puerto Cabezas, Davis, 1902, FAO 1950, Parsons 1955a, Radley 1960, Taylor 1959].

67

3). Monthly rainfall totals, however, do not portray the true situation. Tasbapauni, and most other places on the coast, experience a relatively pronounced dry period with many rainless days from February to mid-May. Along the southern section of the coast (from San Juan del Norte to Monkey Point) there is no marked dry season; the driest month, March, still receives over 5 inches of precipitation but there are more rainless days than at other times of the year. The central coast, from Bluefields to Puerto Cabezas, and the northern coast, from Puerto Cabezas to Cape Gracias a Dios, both have definite dry seasons.

January to late February is a period of intermittent rains, with the number of rainless days increasing until the start of the dry season (*mani*) in February (Table 4). This is the major dry season and lasts

TABLE 4

Distribution of Wet(——) and Dry (---) Seasons over a 4-Year (1966–1969) Period, Bluefields[a]

	Jan.	Feb.	March	April	May	June	July	Aug.	Sept.	Oct.	Nov.	Dec.
			75 days				115 days		16			
1966	├——————+————————┤—+—┤—┤————————+——┤———————————┤
			56 days		11		83 days		19			
1967	├————————————+———————┤—+—+—┤——————————+——┤———————————┤
			84 days		19		107 days		26			
1968	├——————+—————————+—+——┤————————+———+———————┤
			90 days		20		90 days		22			
1969	├——————+—————————+—+——+—————————+——+————————┤
	135 →		76 days	23	12		97 days		21			135 →
Average	├——————+—————————+—+—┤——————————+—+————————┤
			mani (dry season)		*pupu wita* (June rain)		*li mani* (rainy season)		*mani lupia* (Sept. dry)			

[a]Source: U.S. Weather Bureau records, Bluefields.

for some 2½ months, characterized by long periods of rainless days broken occasionally by short-duration showers. Toward the middle of May the eastern skies begin to turn black, and ominous rumblings from the towering thunderheads are heard. The dry season is brought to a quick close when the sheets of lashing rain come, and the rainy season (*li mani* or *kaula pyua ba*) begins. A short dry break (*pupu wita*) in the rains usually occurs around the end of May or the first of June. This generally lasts for 1 or 2 weeks; then rains begin again with increased fury and duration. During July there are many days of endless rain, when the sun is never seen.[6] Charles Bell (1899) who lived on the Miskito Coast for many years, provides an excellent description of what the heavy rainy season is like:

> In June heavy squalls are frequent. . . . Their appearance is dreadfully ominous. Always coming from the east, they rise over the horizon black as night. . . . As they approach, the sky is overcast, the strong trade wind dies away, the wind comes off the land blowing towards the approaching squall, which is seen coming over the lagoon, lashing the waters furiously, and shrouded with a thick curtain of falling rain. First come scattered drops and flying puffs; then the squall breaks wildly over the land, accompanied with torrents as if the clouds had burst. Leaves and branches torn from the trees fill the air. . . . These squalls pass by one after another, sometimes five or six in the twenty-four hours.

> But the great feature of the rainy season is the stupendous rain and thunderstorms which from time to time pass over us. . . . The rain comes down with a roaring sound on the leaves of the forest, like the sound of a great waterfall, or if you are at sea it makes a hissing noise on the surface of the water which is deafening. The lightning and thunder are terrific. I have been since those days over a great part of the world, and never heard thunder to compare with that on the Mosquito Coast [pp. 43–44].

The 3-month period of heavy rains and high winds is followed by a short dry spell (*mani lupia*) lasting from mid-September to early October. There are still heavy thunderstorms, but these are relieved by days of sunshine and calm seas. From October to December the rains can come at any time of the day, with early morning showers and late afternoon storms equally divided.

The differences and alternations between rainy and dry seasons are significant to the occurrence, distribution, and availability of various biota and for the programming and timing of subsistence activities. The Miskito have adopted the 12-month calendar year, but, in actuality, their ordering of time and events still revolves around the seasonal distribution of rainfall and associated environmental changes. The return of the dry season starts a new year in terms of subsistence activities; both the term "dry season" and "year" are expressed by the same Miskito word, *mani* (Conzemius 1932:109). The coming of the dry season signals an end to the long months of rains and strong winds. Available cultivated plant foods are diminished and work must begin to prepare a new swidden site. Turtle are easily taken during the dry season when the sea is calm. With the rains over for a while, turtle meat readily available, and planting to be done, another annual subsistence cycle begins.

---

6. The Miskito use the words *mairin laya* (woman's tears) to describe long periods of drizzling rain that continue without letup.

Within the general pattern of annual precipitation distribution, there are variations in daily rainfall intensity, duration, and occurrence which are significant in the timing of subsistence activities. An individual's decision to go out to sea after turtle, into the forest hunting, to farm his plantation, or to stay home is often influenced by whether rain falls during the night, early morning or day, the number of consecutive days of rain, and the direction and intensity of the winds. If rains are prolonged they can cause widescale flooding along coastal lowlands, isolating animal populations and destroying crops; also, the increased river discharge intensifies and widens the strong littoral current, making coastal traffic impossible for all but powered boats. Most hunting and turtling groups leave in the morning, from about 1:00 to 3:00 A.M. If heavy rain is falling as the men wake up to go they will usually postpone their trip. If rain catches them on the way, no matter, they will go right through it. Most agricultural work is done in the morning from 6:00 to 7:00 A.M. until noon. Again, if it is raining hard the men and women will stay home unless there is no food in the home or to be obtained from kin, in which case they may venture out if their plantation is nearby.

For example, day after day rains are characteristic of July and August weather: "Usually there is a week or so of this weather. The rain generally comes up in the morning with the incoming sea-breeze, and pours in torrents until about four in the evening; then the rain subsides, but a dull slate-blue bank of clouds covers the sky. Then every drenched and hungry living thing comes out to snatch a hasty meal [Bell 1899:45]." Two or three days of continuous rain can be extremely disruptive for food procurement, since little food is kept on hand. I am not talking about continuous showers or back-to-back thunderstorms but a lashing, stinging deluge of heavy wind-driven rain that can numb the body and make the ocean and lagoons impassable.

From an examination of 4 year's daily rainfall records, their amounts, periods, the number of rainless and continuous-rain days for Bluefields, an average seasonal rainfall pattern emerges (Table 5). This pattern appears to be representative for the middle coast, including the principal study area of Tasbapauni. Days having no recorded precipitation or which had almost all of their rainfall during the early evening from 6:00 P.M. to midnight, were called "rainless days." In other words, the important food-procuring hours of the day were rain-free. On the other hand, "continuous-rain" days were designated as having heavy, long lasting showers, usually over 1 inch and lasting 6 hours or more, during these same 18 hours. Many days fell between these two extremes, with rainfall patterns which would not have significantly affected most food-getting activities in Tasbapauni.

TABLE 5

Number of Rainless and Continuous-Rain Days, by Month, 1966 through August 1969, Bluefields[a]

Month	1966		1967		1968		1969		Average	
	RLD[b]	CRD[b]	RLD	CRD	RLD	CRD	RLD	CRD	RLD	CRD
Jan.	13	4	2	2	8	3	9	3	8	3
Feb.	8	2	7	4	17	1	17	0	12	1.5
March	13	1	13	2	18	0	27	0	19	1
April	19	0	12	0	25	0	22	0	20	0
May	13	0	10	4	9	3	16	2	12	2
June	6	7	11	10	8	8	17	5	10	9
July	2	13	2	14	3	15	3	14	2	14
Aug.	4	7	5	6	3	13	10	3	5.5	7
Sept.	9	3	15	5	8	4	—	—	7	4
Oct.	7	5	4	6	11	5	—	—	7	5
Nov.	16	5	6	7	5	5	—	—	9	6
Dec.	13	4	5	5	9	6	—	—	9	5

[a]Source: US Weather Bureau records, Bluefields.
[b]RLD: Rainless days; CRD: Continuous-rain days.

Variations in daily rainfall closely conform to seasonal variation. Thus, during the low rainfall dry season, daily showers are generally infrequent, light and short-lived, while rainy season rainfall is heavy and often of long duration. During the dry season months of March and April, rain-free days occurred, on the average, almost five times as often as they did during the heavy rain months of July and August. In the dry season there is a great deal of activity directed toward subsistence; this is a period of high labor inputs and long-distance journeys. On the other hand, the heavy rain months are a time of food scarcity, reduced subsistence activity, and little or no traveling.

Floods and Coastal Currents

The rivers, full to their brim, rush along with irresistible force, bearing on their red surface great rafts of bamboos, trunks of trees, some with all their leaves upon them, islands of floating grass, piles of plantain and banana trees, and sometimes dead deer and waree; all the lowlands near the sea are flooded over.

—Charles Napier Bell (1862:249)

Important effects on the distribution and types of fauna, and on hunting, fishing and agriculture are wrought by heavy flooding and associated strong coastal currents. Along the upper and middle reaches of the rivers, water levels can rise 20 feet or more after a prolonged period of heavy rainfall. High water floods during June, July, and August, which occur along all major Miskito Coast rivers with good size catchment

basins, can have drastic effects on riverbank agriculture. Entire plantations are often washed away during major floods. Near the coastline,
where river courses widen and riverbanks and natural levees are low,
flood waters spill over onto the floodplains causing water levels of coastal
lagoons, rivers, and creeks to rise until large sections of the lowland
are continuous sheets of water; the whole relationship of water to land
is thereby rearranged. Here too, plantations are flooded, crops
destroyed.[7] Also, with few high places to seek refuge, animal populations
are isolated and many drown. Often during flood times hunters will
go along rivers looking for deer trapped on hillocks and natural levees.
Then large kills are made.

While on a journey up the Río Prinsapolka, Bell (1899) commented
on the heavy floods that had turned the area into a huge raging body
of water:

> The flood rose during the night, and when I woke I hardly recognized the place.
> I had known the village as 60 feet above the river; now it was not more than 10
> feet. On the opposite side the banks were low, and the river was the colour of clay,
> and was bearing along the most astonishing quantities of stuff. Among immense
> rafts of trees, grass, and bamboos we occasionally saw canoes, thatched roofs, and
> piles of banana-trees. Many snakes were seen swimming about the river, and dead
> deer and peccary floated past.
> The bush was too much flooded for hunting, and the creeks for fishing [pp. 245–246].

For coastal Miskito, the direction, intensity, and width of littoral currents are major determinants of whether or not they will be able to
use their dories to go out after turtle or, as on the upper coast, whether
or not they can get to market in Puerto Cabezas. Along with wind direction and intensity, currents are a major topic of conversation among
the Miskito.

Periods of heavy littoral currents are associated with the June to July
wet season when northeast trade winds push the currents southward.
Large rivers that drain the highlands of Nicaragua and the coastal lowlands pour out vast quantities of fresh water to the sea. These waters
are carried southward along the coast by prevailing winds and create
a belt of red–brown fresh water, 8–10, sometimes 20 miles off the coast.
During August when heavy rains start to abate and clear days begin,
it is often impossible for turtlemen to go out to sea because they cannot

7. In July 1971 prolonged heavy rains led to widespread flooding and destruction of riverside
agricultural sites along the Río Kuringwas and Río Wawashan west of Tasbapauni. Tasbapauni villagers
said that this was the worst flood they could remember. Raging floodwaters stripped away vegetation,
caved in riverbanks and washed large tangles of debris into Pearl Lagoon. Tasbapauni agricultural
grounds along the lagoon were covered with water and large losses of crops resulted. Heavy floods,
such as this one, are known as "top gallon floods."

paddle against the heavy current. There is usually a lag time in August and September for rivers to go down and currents to slack off. The belt of fresh water also influences distribution of close-in fish and shrimp. Many saltwater fish and shrimp are not fresh water tolerant and are driven off by the littoral flood waters.

Winds

Some days the heating of the shore by day and the cooling by night gave you a fitful daily seabreeze – landbreeze cycle, with lines of thunderstorms where the wet doldrum air soared at the land-edge. Some days the trade wind came back down and held strong for awhile; and all the dugouts sailed madly for where they had to go before it should leave again.

—Archie Carr, in THE WINDWARD ROAD (p. 132)

The everyday life of the coastal Miskito depends much on their sailing dories; consequently, the direction and intensity of the wind is of great importance. Similarly, the persistence of the trade winds and the occasional hurricanes play significant roles in the ecology of the area. The Miskito have at least 25 words and terms for winds and wind conditions, which gives some idea of how acutely aware they are of wind variations.

Most notable features of the wind patterns along the east coast of Nicaragua are the dominance of the Northeast Trades and the low frequency of periods of calm. As shown in Table 6, winds from the northeast sector blow 48% of the time, while calms occur only 2% of the time. In January northeast winds blow hardest, whipping up seas and lagoons and making water transport difficult. During the February through May dry season, the trades gradually diminish in intensity and frequency. By June average wind intensities begin to increase and violent squalls and thunderstorms come out of the east. In July, the rainiest month, northeast and east winds continue to persist and some of the highest wind velocities are recorded. Along with high river discharges, the northeast winds create strong currents. The general weather pattern of July continues in August but intensities decline. September is a period of relative calm coinciding with the secondary dry season. In addition, wind direction becomes more variable and inconsistent, making sailing unpredictable. But September is "pretty weather," and the calm is extremely important for food procurement after months of strong winds and heavy rains.

During October wind direction becomes the most variable and contrary of any month in the year. This is the time when turtlemen, trying to get out to, or back from the offshore turtle banks, have the most difficulty.

TABLE 6

Recorded Wind Observations off the East Coast of Nicaragua, 1884–1934[a]

Month	No. of obser- vations	Mean velocity (mph)	Percentage of observations from								
			N	NE	E	SE	S	SW	W	NW	Calm
Jan.	1060	12.5	15	61	17	1	1	1	1	3	0
Feb.	952	11.1	17	55	19	3	1	1	1	2	1
March	1062	9.7	13	56	20	4	1	1	1	2	2
April	983	9.3	16	51	23	4	0	1	1	2	3
May	1038	8.4	12	44	26	6	3	1	2	3	3
June	1331	9.3	6	40	33	6	2	2	3	4	4
July	1338	11.6	7	54	29	3	1	1	1	2	2
Aug.	1303	9.3	9	54	22	4	1	1	4	3	3
Sept.	1235	6.5	12	33	26	8	3	3	4	5	6
Oct.	1368	7.2	14	28	18	7	5	5	8	10	5
Nov.	1241	9.3	23	38	13	4	1	3	9	7	2
Dec.	996	12.0	15	60	18	2	0	1	1	2	1
Total:	13,907										
Mean		9.7	13	48	22	4	2	2	3	4	2

[a]Recordings are for Greenwich noon time, which would be about 6:00 a.m. in Nicaragua. Source: US Hydrographic Office 1948:73.

A 15–20-mile paddle against a strong wind is hard work and cuts down on the frequency of turtling expeditions and on the use of dories for hunting and visits to the farms. By November strong *nortes* blow, lasting through December. When a *norte* is on, no one can go out in a *dori* unless it is within some of the inside lagoons and rivers sheltered from the fury of the north wind.

Winds are major factors in the ecology of the area and of Miskito subsistence. The Miskito have adapted many of their food-getting pursuits and schedules to coincide with favorable wind conditions. Months with especially unfavorable winds are usually referred to in Miskito by the direction or type of wind. For example, July is often called *pastara kati* or "strong wind month"; October is referred to as either *saut kati* (south wind month), or *prari kati* (hurricane month); November is always known as *yahbra kati* or "north wind month."

The east coast of Nicaragua lies outside the principal hurricane zone, and although hurricanes are infrequent, they are known to have caused widespread ecological alterations. Their suspected influence on the wild biota is thought to be severe, but the extent is unknown. The effect of hurricanes on marine and terrestrial fauna has hardly been studied.[8]

8. Stoddart (1962, 1963, 1965, 1969) and Vermeer (1963) present observations on the effects of hurricanes on the cays of British Honduras, including some information on marine fauna.

It can only be inferred that there are serious and widespread disruptions in animal populations resulting from initial deaths caused by high winds and changes in food chains. Pim and Seemann (1869:372–377) related a vivid account of an 1865 hurricane that hit the Bluefields area, causing large-scale destruction and driving some large predators such as jaguars into Bluefields in search of food.[9] Hurricanes also cause extensive blow-downs of subsistence crops, as well as in commercial banana and lumbering areas. Hurricane blow-downs are thought to be the reason for unusual concentrations of secondary plant species in areas of tropical forests of the mid-Miskito Coast (Carr 1953:160–161).

While being relatively rare, the few hurricanes that have passed over the Miskito Coast have probably more seriously disrupted the local food system of the Miskito than any other short-term environmental factor such as unusually heavy rains or fires.

From 1865 to 1971 there were at least 28 hurricanes and tropical storms that hit the coast and adjacent shallow coastal waters of Nicaragua (Table 7). The inhabitants of Tasbapauni still talk about the hurricane of 1906 and how it completely leveled all the beach plantations and forest, destroyed most of their crops, coconut trees, and caused many of the game animals to disappear. This hurricane was so severe that older villagers date their age from its occurrence. Today any rumor of a hurricane off the coast of Nicaragua will bring all the people back to the villages.

The Miskito attribute the opening and closing of many river bars in the past to hurricanes. Any change in local hydrographic conditions accompanied by a change in water salinites will result in a change in aquatic fauna. For example, oyster beds will die out if a bar is closed and they are deprived of circulating salt water. Tasbapauni Miskito claim that Karaslaya Bar (once opposite the mouth of Kuringwas River) was closed by the 1906 hurricane and that the extensive oyster beds perished.

Most hurricanes and tropical storms occur along the Miskito Coast between September and November, particularly in October. This is espe-

---

9. The hurricane hit the mid-Miskito Coast between Wounta Haulover and the southern end of Bluefields Lagoon, a distance of some 90 miles, and caused damage from Corn Islands to about 30 miles inland. In referring to the Bluefields area it was observed:

> The lagoon was covered with trees, branches, and leaves for a long time; the water turned quite black, and the fish died by the hundreds. no doubt poisoned. They floated on the surface of the lagoon until the exhalations arising from their dead bodies became almost unbearable. . . .The beach all round was lined with hundreds of dead fish, alligators, sharks, and a variety of strange shells, while on the river banks there were great quantities of dead maniti.
> A few days later the parrots came to the settlement in thousands, and great numbers dropped dead from sheer starvation. Then, the tigers [jaguars] made their appearance, lean, gaunt, and savage, eating up everything that came their way . . . no less than eight were killed in one day [Pim and Seemann 1869:375–377].

TABLE 7

Hurricanes and Tropical Storms That Have Hit the Miskito Coast of Nicaragua, 1865–1971[a]

No.	Date			Area hit
1	Oct. 18,	1865	(H)[b]	Bluefields–Pearl Lagoon
2	Oct. 5,	1882	(H)	Miskito Cays
3	Oct. 10–11,	1892	(H)	Miskito Cays, Cape Gracias a Dios
4	July 5,	1893	(H)	Miskito Cays, Cape Gracias a Dios
5	Sept. 21,	1898	(TS)	Miskito Cays
6	Sept. 22,	1901	(TS)	Miskito Cays
7	June 10,	1902	(TS)	Miskito Cays
8	Oct. 9,	1906	(H)	Pearl Lagoon, Tasbapauni
9	Oct. 17,	1908	(H)	Prinsapolka
10	Sept. 9–10,	1911	(H)	Río Grande
11	June 22–23,	1913	(H)	Sandy Bay
12	June 30,	1916	(TS)	Miskito Cays
13	Nov. 12,	1916	(TS)	Miskito Cays, Cape Gracias a Dios
14	Sept. 17,	1920	(TS)	Cape Gracias a Dios
15	Oct. 17,	1926	(TS)	Miskito Cays
16	Nov. 13,	1926	(TS)	Miskito Cays
17	Sept. 28	1933	(TS)	Miskito Cays
18	Nov. 16,	1933	(TS)	Bluefields
19	Oct. 25,	1935	(H)	Mid-Río Coco
20	Sept. 19,	1940	(TS)	Prinsapolka
21	Oct. 23,	1940	(TS)	Prinsapolka
22	Sept. 27,	1941	(H)	Old Cape, Cape Gracias a Dios
23	Aug. 29,	1945	(TS)	Miskito Cays
24	Nov. 3,	1949	(TS)	Puerto Cabezas
25	May 27–28,	1953	(TS)	Northeast Nicaragua, Prinsapolka
26	July 13,	1960	(TS)	Miskito Cays
27	Sept. 9–10,	1971	(H)	Old Cape, Cape Gracias a Dios, Lower Río Coco
28	Sept. 18–19,	1971	(H)	Monkey Point

[a]Compiled from Cry (1965), and Pim and Seemann (1869:372).

[b]H is hurricane, having winds, 74 miles per hour plus; TS is tropical storm, having winds, 39–73 miles per hour.

cially important because for many crops this is harvest time: rice, the first cassava, and other ground provisions. It is the critical period after the long hunger of the wet season. This is also the time when meat from fishing and hunting again becomes important in the diet, and when market crops are sold. In the face of a hurricane all work stops; if a hurricane does occur, the effects are more than just the loss of that year's plantation crops, tragic enough in itself, but also the loss of coconut groves, which take 5–10 years to replace; and probably an equally disastrous, but unknown effect on fish and game animals.[10]

Monthly wind patterns and occasional hurricanes, then are major factors in the ecology of the coast and of the Miskito. Direction and intensity of winds, and their seasonality greatly affect movement by *dori* to hunting, fishing, and agricultural areas with open exposure to the sea. Prevailing northeast trade winds often hamper mainland-to-offshore sailing, and during the months of June through August, the trades intensify the littoral current of fresh water that influences not only the occurrence of different fish species but also the Miskito's accessibility to the turtle banks. The constant winds, predominantly from the north to east quadrants, effectively cool windward locations, making work much more bearable in the hot, humid climate. The windward location of many Miskito villages is undoubtedly partially influenced by the cooling exposure to the prevailing winds. When sea winds blow for a long period and over the long stretch of open sea, very rough surf can break on exposed beaches, along which many villages are located. This makes getting a *dori* "outside" hazardous. But the Miskito are so skillful that dories rarely capsize, even in the roughest surf. Sea winds also can cause open bars to be very rough with breaking waves and white water all about. Then, even the best boatmen will not go out.

Temperature

Available temperature and relative humidity data are extremely scarce for the coast and are so short-termed as to be of little value other than indicating typical mean monthly temperatures.[11]

In Table 8, temperature figures for Bluefields are given. Average monthly temperatures usually range from 75 to 80°F, while mean relative humidity readings commonly vary from 78% to 90%. Along the coast maximum temperatures occur during May and June, and in September

10. In September 1971, two severe hurricanes hit eastern Nicaragua causing widespread loss of life, damage, and flooding. Hurricane Edith slammed into the northeastern coast, wiping out the Miskito village of Old Cape, greatly damaging 20 other villages and temporarily displacing 3000–4000 Miskito along the upper coast and lower Río Coco. Agricultural fields were leveled by winds and inundated by wind-driven water. Hurricane Irene passed to the south, crossing Nicaragua after damaging the Monkey Point area south of Bluefields. Heavy rains, winds, and floods were experienced all along the coast. In Tasbapauni, large amounts of crops were ruined in the exposed plantations along the beach.

11. Bennett (1967:11–12) has commented on the poor nature of temperature and atmospheric humidity records for Middle America in general. For ecologically significant temperature and humidity measurement he suggested recording thermometers and thermistor thermometers to be set up in plant and animal habitats, including those of man, over long periods of time.

TABLE 8

Temperature, Bluefields, 1968[a]

	Jan.	Feb.	March	April	May	June	July	Aug.	Sept.	Oct.	Nov.	Dec.
High	83.11	83.27	83.61	84.73	85.86	87.10	83.48	84.19	86.56	86.29	83.76	84.16
Low	71.06	68.86	71.06	72.86	74.41	72.73	73.09	72.22	70.63	69.61	68.60	67.58
Mean	77.11	76.06	77.23	78.79	80.18	79.91	78.28	78.20	78.59	77.95	76.23	75.87
Variation	12.05	14.41	12.55	11.87	11.55	14.39	10.39	11.97	15.93	16.68	15.06	16.58

[a]Average highs, lows, and daily variation, in degrees Fahrenheit. Source: US Weather Bureau records, Bluefields.

and October; minimum temperatures are recorded from early November through February. Dry season temperature fluctuations are low, while absolute temperatures gradually increase to their mean high in May. At the start of the rainy season, relative humidity and daily temperature fluctuations increase, while mean temperatures decline due to the cooling effect of daily rains. High maximum temperatures are experienced in September, but rains continue to hold down the means. From September through December cold north winds and chilling showers, broken by days of beautiful clear skies, account for the 15 to 16°F monthly variation in temperature.

Averages and means do not portray true temperature extremes that often occur along the coast. There are many sudden temperature changes similar to those between dry and wet seasons during the November through February period; then cool temperatures and occasional cold *norte* winds are experienced. A July squall can drop temperatures 10°F in a few minutes. The Miskito attribute the increase in instances of people with fever and colds during July and August, and between December through February, to such sudden changes in temperature. This also reduces the overall number of possible occasions for food-getting.

Based on the inadequate temperature data it is impossible to make any comments on any possible effects that such a narrow range of temperature fluctuations might have on other aspects of the environment. To be sure, the relatively constant temperature range permits a wide variety of crop choices, and should not be detrimental to faunal or floral patterns. However, sudden drops in water temperature, usually associated with *nortes*, if severe enough, can indeed cause high fish mortality in shallow waters (Collier 1964:141–142; Storr 1964:31).

Extremely high noonday temperatures during much of the year, coupled with high humidity, makes agricultural work in the plantations very unpleasant after about 11:00 A.M. Even in plantations along the beach, midday temperatures are oppressive once one leaves the beach itself, and goes behind the beach scrub. Hard work, such as using a machete in the open fields, is just too taxing by noontime. Consequently, in order to avoid the heat, as well as late afternoon rain showers, villagers usually leave for their plantations between 2:00 and 6:00 A.M. depending on the distance. Work usually stops by noon. With respect to high temperatures at noon, Bell (1862) remarked poetically: "All Nature seems to retire to rest for a season when the sun, having reached his highest point, sends down a flood of light and drowsy heat [p. 246]." Because most hunting takes place in the shade of swamps and forests, and because turtle fishing is out on the sea where there is a breeze, these activities are not so much affected by the high temperatures and humidity.

Relatively constant high temperatures and high humidity are also conducive to microorganic activity, and running sores. Surface infections are a constant problem with the Miskito. The high temperature–humidity factor also hampers food preservation and storage. Unless eaten quickly or smoked, meat spoils rapidly, as do shrimp and fish.

Summary

Daily as well as seasonal changes in wind direction and intensity, coastal currents, rainfall and, to some extent, temperature fluctuations, are significant in the ecology of coastal Miskito food procurement patterns and for the resource accessibility and mobility of the people. Throughout the discussion on climate it has been emphasized that it is often the microvariations that can be extremely important in the subsistence system. Even in the humid tropics where weather and climatic patterns are often mistakingly stereotyped as consistent and monotonous, daily and seasonal variations of weather elements play major roles in the ecological complexities of the food quest.

The Miskito pay close heed to daily weather. The annual march of the weather, its major and minor patterns, are significant parts of the Miskito's conception of time, place, and season. All the wet and dry seasons, south wind and north wind months, bad current months, and "sweet weather" months, fit into their overall time system within which the form and relationship of the components and activities of the subsistence system operate.

If by chance, a Miskito were to wake up Rip van Winkle-style, not knowing the time of year, I dare say that after a look at the wind direction, height of sun, sea condition and color, presence or absence of rain clouds, he could take you straightaway into the forest to gather a wild fruit that ripens in only that month.

BIOPHYSICAL PATTERNS

Life reaches its greatest diversity in tropical seas and tropical forests. Warmth, light, moisture, the three essentials for life, are here always present and dependable.
—Marston Bates, in THE FOREST AND THE SEA (p. 24)

The biophysical patterns of the Miskito Coast present radically different environmental situations with which the Miskito must cope in procuring food. For description purposes, the coast can be roughly divided into four large-scale ecosystems representing broad general assemblages of biological communities and environments: (1) the tropical rain forest; (2) the pine savanna; (3) the coastal complex of beach, lagoon, and swamp; and (4) the coastal offshore waters, coral cays, and reefs (Fig. 15). Within each of these broad ecosystems are myriad microenvironments and species which the Miskito specifically exploit by hunting, fishing, agriculture, and gathering.

In their description and categorization of plants and animals, the Miskito recognize small variations in form and appearance and have separate terms for distinct species. In the rain forest, for example, they are often able to determine tree species without the aid of leaves, seeds, or blossoms, all of which are many feet overhead. Differentiation of tree species is aided by a consideration of bark color and texture and sometimes by a machete cut to reveal inner bark color and a taste sample of sap. They perceive taxonomic relationships and include many of these in their nomenclature. In general, nondomesticated plants and animals not important for food or resource materials are designated usually by species, sometimes generic taxon. Wild and domesticated biota used in subsistence and livelihood, on the other hand, are often distinguished in minute detail, at the level of subspecies or even varieties.

Tropical Rain Forest

One steps through the wall of the tropic forest, as Alice stepped through the looking glass; a few steps and the wall closes behind. The first impression is of the dark, soft atmosphere, an atmosphere which might be described as "hanging," for in the great tangle of leaves and fronds and boles it is difficult to perceive any one plant as a unit: there are only these hanging shapes draped by lianas in the heavy air, as if they had lost contact with the earth.
—Peter Matthiessen, in THE CLOUD FOREST (pp. 44–45)

"The mature tropical rain forest is probably the most intricate, productive, efficient and stable ecosystem that has ever evolved [Rappaport 1971:117]." The energy produced by the system is distributed among

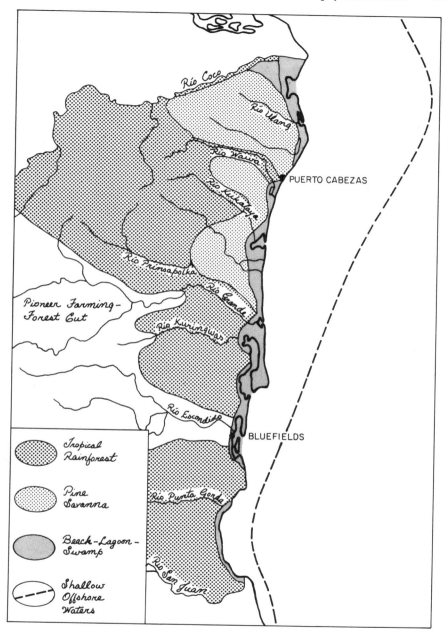

Fig. 15. Major ecosystems of eastern Nicaragua.

a large variety of species, each of which is made up of a relatively small number of individuals (Geertz 1963:16). The rain forest, then, is a very generalized, highly diverse ecosystem, with a tremendous variety of biota. Energy and nutrient sources are stored in the forest biomass and circulation, recycling, and recovery rapidly occur within a closed cycle. The continuous dense, storied canopy cover captures and stores energy and nutrients and protects the thin soils from erosion and excessive temperatures from solar radiation. Maximum capture and utilization of energy and nutrients is facilitated by the diversity of species of biota, each with different and overlapping requirements (Meggers 1971:17–18).

The tropical rain forest of the Miskito Coast is one of the last remaining extensive stretches of this type of vegetation left in Central America. The rain forest is made up of three to four stories of foliage, highly dispersed broadleaf evergreen species and an unbroken canopy cover. Associated are unusually high concentrations of palm communities along some river courses and floodplain areas; gallery forests extending right to the coastline through areas of savanna, marsh, and swamp; and secondary forests characterized by dense underbrush and dominance of pioneer species[12] (Fig. 16).

Fig. 16. A Miskito hunter entering the tropical rain forest.

12. The various components of the Miskito rain forest are discussed by Taylor (1963) and Radley (1960), each giving long lists of common tree species.

Along the flanks of the interior mountains are areas of lower montane rain forest in the upper Río Bocay region and extending to mining areas around Siuna and Bonanza. A belt of seasonal mixed rain forest occurs in the headwater portions of the Río Grande, Río Siquia, and south to the Río San Juan (Taylor 1963:33). The coastal plain and intervening hilly areas are covered with lowland rain forest, extending up to the pine savanna in the north and thickly cloaking the middle and lower Miskito Coast. The rain forest penetrates the savannas along rivers in the form of gallery forests, especially along the Río Coco and Río Wawa. Lowland tropical rain forest covers an area of coastal plain which is low-lying, few spots having elevations of more than 150 feet. The terrain is gently rolling, broken by occasional low hills and dissected by many rivers and creeks.

It is difficult to designate the most common tree species, since their distribution is so highly dispersed. Valuable timber trees include mahogany (*Swietenia*), cedar (*Cedrela*), and cedro macho (*Carapa*). These all occur in the coastal rain forest and have supported lumbering concerns for many years. To be sure, these trees are valuable to the Miskito for making dories and pitpans.

Many of the favorite game animals of the Miskito are found in the rain forest. These animals do not range evenly throughout the rain forest but are concentrated in areas with favorable habitats. The Miskito do little hunting in the deep selva where game is scarce. Most of their hunting efforts are directed toward white-lipped peccary (*Tayassu pecari*), white-tailed deer (*Odocoileus virginiana*), and occasionally brocket deer (*Mazama* sp.) and tapir (*Tapirella* sp.). The Miskito take many other animals for food such as birds and small mammals.[13] Recently, high market prices for skins from jaguars (*Felis onca*), ocelots (*F. pardalis*), and margays (*F. wiedii*) have prompted the Miskito to hunt more in the rain forest, where they may stay for a month or more setting deadfall traps.

Gallery Forest and Floodplain. The many rivers of the Miskito coast have permitted easy access to rain forest areas for both coastal and riverine inhabitants. The Kuringwas, Río Grande, Prinsapolka, Kukalaya, and Wawa are major water routes into the rain forest for the Miskito of the middle coast area. North coast Miskito and Río Coco Miskito are generally confined to the gallery rain forests along river courses for hunting. The rain forest, however, extends down to the Río Coco west of Leimus. Rivers have long facilitated extraction of materi-

---

13. See Appendix B for a list of fish and game animals taken by the Miskito.

als from the rain forest by the Miskito, and also by foreign mining, lumber, and banana companies.

The long rivers have allowed the Miskito to concentrate their plantation sites along the natural levees and floodplains in alluvial soils, rather than away from the rivers in interfluvial rain forest areas where more difficult agricultural conditions are found (poor soils, problems of clearing off vegetation, and carrying out crops). Miskito will go long distances to locate a suitable site along a river or creek to plant their crops. Travel distances of 20–30 miles to agricultural areas are common for coastal people.

The tree-walled rivers and streams that cross the coastal plain are usually dominated by tropical rain forest biota. Often the vegetation shows only small differences from rain forest species. Usually there are more local dominants than in the rain forest, and a greater predominance of palms. The riverside site is one of the most important areas for settlement and food-getting on the Miskito Coast. For example, more than one-half of the total Miskito population of Nicaragua resides along the Río Coco.

The Miskito are above all a water people. Whether it be river, lagoon, or sea, the Miskito live by and travel by water. Within the rain forest, it is the riverine site that is important for Miskito transportation, settlement, agriculture, and of course, fishing. The rivers permit access to favorite hunting, gathering, and agricultural areas. Sauer (1958), in discussing man in the ecology of tropical America, made a point on the riparian habitat that could be well directed to the Miskito:

> However forbidding the unbroken tropical forest was to the most primitive folk, the water breaks in the forest invited his entry by the multitude of streams that keep their sunlit way open through the forest. Lake and seashore also offer open spaces. In large measure primitive man is a riparian creature anywhere; he moved into the tropical forest along the edges of the water, lived by the water, and there gathered everything needful. It made little difference how great or tall the forest, the sunny avenues of water provided a congenial environment; widely branching, they led far inland toward the watersheds. The riparian habitat, I wish to submit, has been favorable to progress, since its environment is diversified as to plant and animal life both of land and waters [p. 106].

The distribution of Miskito populations alongside aquatic environments is significant not only in terms of the ease of mobility by water transport, but of the concentration of diverse ecological zones and biota within compact spatial units. The heterogeneous ecological makeup of juxtaposed land and water environments, facilitates advantageous playoffs in the exploitation of aquatic and terrestrial fauna, enabling, as we shall see, a more consistent and dependable supply of animal protein. According to several studies, the aboriginal settlement patterns and high population densities along tropical rivers, lagoons, and sea-coasts, may have been determined more by the avilability of protein-rich fish and game resources than by agricultural possibilities alone (Denevan 1966, 1970; Lathrap 1968; Carneiro 1970).

In addition to their importance as food-getting and settlement sites, gallery forest areas have been of important zoogeographic significance as mesic corridors for tropical rain forest biota (Carr 1950:588). Many tropical forest animals such as jaguar, ocelot, tapir, and collared peccary (*Pecari tajacu*) occur in beach forests and swamps, far removed from the tropical rain forests by intervening lagoons, mangrove and palm swamps, with only gallery forest providing a possible dispersal route.

Most of the riverine areas, with the exception of the extensive palm forests, have been long utilized for agriculture; therefore, little mature forest is left. Taylor (1959:81) mentioned the following trees as being common on the Río Coco floodplain, the single most important riverine site for the Miskito: silk-cotton tree (*Cebia pentandra*), *Croton glaballus, Dialium guianense,* canal (*Nectandra globosa*), guayavo negro (*Terminalia amazonia*), guásimo colorado (*Luehea seemannii*), and *Erthrina glauca.*

On many of the lower river courses where low-lying lands are wet most of the time, large areas are frequently covered with palm forests of several varieties: hone (*Coroza oleifera*), cohune (*Attelea cohune*), papta (*Acoelorrhaphe*), and several other genera including *Bactris* and *Thrinax.*

Pine Savanna

From one of the heights . . . we had a most extensive view of the country, which as far as the eye can reach, is all low savannah land, covered with coarse long grass, and occasional pine ridges, with remarkably large, and fine timber.

—Orlando Roberts (1827:127)

Although the pine savanna is one of the largest ecosystems in area on the Miskito Coast, the Miskito rarely exploit it for food resources. The soils are too poor to support subsistence agriculture, little wild game is found on the savannas, nor is there much in the way of wild fruits to be gathered. Some cattle are grazed on the savanna grasses, however. The low food potentiality of the savannas constricts the Miskito's food-getting activities to the rivers that cross and border the savannas (Fig. 17).

The savannas present a golf-course-like appearance of pine and grass-covered rolling topography, with scatterings of pine ridges and clumps of papta palms in the humid depressions. Along the road from Puerto

Fig. 17. Gallery forest on the Río Likus near Sisin, with pine savanna extending to the north and south. Composed principally of rain forest flora, the gallery formation is the major focus of the subsistence farming and hunting for the Miskito villages on the savanna.

Cabezas to Waspam–Bilwaskarma, the savannas stretch away endlessly until the yellow-green of the grasses and pines merge to dark gray on the distant horizon. It was an amazingly different feeling to be on the open savannas where space and horizons exist, after months of fieldwork in an area of tropical rain forest. In the rain forest one's conception of space is confined to the vertical, from ground level to the canopy cover, where a chance glimpse of blue sky is a rare event. Visible horizontal distances are measured in yards, not miles as on the savanna. In this respect the savannas evoke a feeling of freshness and openness similar to being at sea in a *dori*, with a steady breeze and an occasional look at the beach from the crest of a rolling sea. But the similarity ends there. For the Miskito the savannas are barren while the sea is rich in food resources.

Extending from Cabo Camarón in Honduras south to the Río Grande, the pine–grass savanna is located between the tropical rain forest to the west and the complex of lagoon, beach, and swamp to the east. The savanna continues beyond the Río Grande in isolated patches on raised beach ridges to a point just north of Bluefields (Radley 1960:161).[14] These are the southernmost naturally occurring stands of pines (*Pinus*) in the New World (Parsons 1955a:36).

The savanna occurs in two separate divisions: the Río Coco–Río Wawa savanna, and the Alamikamba–Limbaika savanna. These two are separated by a broad belt of tropical gallery rain forest along the Río Wawa and extensive stands of palm swamp. Even within the savannas occur many areas of swamp, marsh, islands of hardwoods, and belts of gallery forests which may occupy one-third of the general pine savanna area (Parsons 1955a:38). Upland portions of the savannas are covered with open stands of Caribbean pine (*Pinus caribaea*) and stunted pyro-phytic trees such as sandpaper trees (*Curatella americana*) and nance (*Byrsonima crassifolia*), with a scattered bush cover and an extensive ground cover of grasses and sedges (Taylor 1959:77; 1963:48). Low-lying wet areas have a low herbaceous cover and conspicuous pockets of palms, mainly papta (*Acoelorrhaphe wrightii*).

In contrast to the tropical rain forest, the savannas are almost devoid of wildlife.[15] The Miskito do occasionally hunt white-tailed deer and may catch an armadillo (*Dasypus* sp.), but most hunting and fishing activities are focused on tropical forests, rivers, and the sea.

Miskito savanna settlements are located along the edges of the savannas near beach–lagoon or riverine areas. The focus of village life and food systems is on the coast or in the gallery forests, not on savanna, although Miskito men did work for many years for lumber companies cutting pine trees for export lumber. In terms of the historical role of the savannas to Miskito livelihood, Parsons (1955a) remarked:

> It seems highly improbable that the Miskito savanna surfaces were ever farmed exten-sively, at least in their present highly leached state. Moreover, the native Indians

14. Various explanations for the occurrence of pine savanna in a humid tropical environment have been offered. Taylor (1963:48) suggested that the pine savannas are a fire-caused disclimax from what would be rain forest or swamp forest in a climax state. Radley (1960:177–179), on the other hand, postulated a possible preman origin of the savannas resulting from an influence of the variable length of the dry season and edaphic controls. He deemphasized the significance of a fire-induced savanna but did not preclude the possibility. Parsons, in what appears to be the first detailed investigation on the Nicaragua–Honduras savannas, argued convincingly for a more multifaceted origin and mainte-nance of the savannas based on long burning of the area, suppressing a possible former xerophytic broadleaf forest, augmented by occasional blow-downs of broadleaf by hurricanes. Parsons (1955a:44–48) suggested that the factor of poor drainage on the savannas due to impermeable clay–gravel horizons may have been a result of the burning and change in vegetation cover rather than a cause.

15. The vast extent of the savannas, over 250 miles in length and varying from 30 to 100 miles in width, must be an important deterrent to tropical forest faunal movement. Since the date of the origin of the savannas is not known, little can be postulated about their effect on past distribution patterns of Neotropical fauna in the predominantly tropical forested Caribbean lowlands.

of the region were at best casual farmers who took their living chiefly from the seas and the hunt. Although their cultivations have probably always been restricted to the narrow strips of alluvium along the streams and behind the coastal beach ridges, it appears that they have habitually burned the savannas for as long as anyone can remember, whether to aid in hunting, to improve grazing, or simply for excitement [pp. 45–46].

Bell (1862) gave support to Parsons' thesis that the Miskito regularly burned the savannas when he noted that in April "the Indians set fire to the savannahs; and if the wind happens to be north, the whole country is obscured with smoke, the sun becomes as red as in an eclipse, and the smell of the fire is perceived for hundreds of miles [p. 247]."

Along the Río Coco, especially between Waspam and Living Creek, most of the villages are located on the elevated savanna edge, away from the low-lying hot, humid floodplain. This is probably because of occasional severe floods, and the presence of bothersome insects along the rivers, whereas the savannas are markedly cooler, being exposed to the trade winds, and because communication between villages is easy across the savannas.

Besides lacking food resources, the savannas have acted as a barrier between the two major Miskito areas and lifeways: the riverine Río Coco Miskito, and the sea-going coastal Miskito.

Coastal Beach, Lagoon, and Swamp

The longer I dwelt on Inagua the more certain it became that the day's activities would terminate within sound or reach of the surf. When the inner jungle became too hot to be tolerated and seemed devoid of any life ... it was in sheer relief that I turned to the edge of the sea. Here always was life and movement, cool air and activity.
—Gilbert Klingel, in THE OCEAN ISLAND (p. 112)

This is one of the most complex ecological zones of the Miskito Coast and is the second most important locale in terms of settlement. The dominant ecological controls on beaches, coastal lagoons, and swamps are both terrestrial and marine. Located between the rain forest and savannas to the west and the offshore marine waters to the east, this is a very dynamic ecological area. Convergence of land–sea factors gives character and unity to this ecosystem. Transition from land to sea is not an abrupt break occurring at the strandline. A great deal of overlap of marine and terrestrial environmental influences and biota occur within the low-lying beach–lagoon–swamp areas fronting the sea. The littoral current, the onshore winds, the influx of sea water and organisms into lagoons and up the rivers, are among some of the important marine influences; seasonal river flooding, the land wind, and occurrence of rain forest biota are some of the dynamic factors originating from the land sector.

Living within the narrow band of beach–lagoon–swamp complex, the Miskito are able to exploit both land- and sea-based food resources. To do this they have to have the knowledge and ability to utilize extremely different environments, each with a complex ecology.

This ecosystem contains coastal lagoons, estuaries, palm and mangrove swamps, marshes, river bars, and beaches (Fig. 18). Here is the contact zone between fresh water from the land and salt water from the Caribbean. For the Miskito it is a water world where the relationships between land and water and between fresh water and salt water change constantly throughout the year.

The shore is composed of long stretches of smooth sandy beaches, broken only by headlands at Monkey Point south of Bluefields, at El Bluff across from Bluefields, and at Bragman's Bluff near Puerto Cabezas. The shoreline is made up of a series of spits, bars, and north–

Fig. 18. Coastal lowland near the Río Wawa, September 1969. A complex of lagoons, rivers, swamps, and marshes, this area turns to a vast sheet of water during the rainy season.

south parallel beach ridges. The beach zone is usually backed by numerous lagoons and swamp areas where former estuaries have been cut off by advancing spits (Radley 1960:96). There are several very large lagoons separated from the sea by only narrow haulovers. All of the large lagoons are fed by one or more major rivers and have shallow bars giving access to the sea. Often surrounding the lagoons and dispersed between them are marshes, and mangrove and palm swamps. The entire coastal area is extremely flat, with most higher elevations consisting of 2- to 6-foot beach ridges and isolated 10- to 15-foot hillocks.

The lagoon–beach system is of varying width. In some places, where the savanna hugs the beach, it is only a few hundred feet wide; in other areas backed by large lagoons and vast swamps the zone may be 10 or more miles wide. The most extensive beach–lagoon areas are found around Old Cape and Sandy Bay, Pahara Lagoon, Wounta Lagoon, and Bluefields Lagoon. With the exception of Bluefields Lagoon, all of the others are important centers of coastal Miskito settlement.

Three of the largest rivers on the coast (the Río Grande, Prinsapolka, and Río Coco) enter the sea directly and have large deltas. Most of the remainder of large rivers empty into lagoons and offer easy access to interior areas for coastal inhabitants. These lagoons have shallow bars which can usually be crossed with dories even in rough weather. On the other hand, the river bars can be very dangerous during flood time because of strong outgoing currents, breaking waves, sharks, and shoals.

The network of coastal lagoons, rivers, and creeks is interconnected for long distances providing a calm inland north–south water route when the sea is too rough for travel. This inland water system also enables the Miskito to exploit far distant food resources and agricultural areas. It is possible to go from Bluefields to the Kuringwas River, located at the northern end of Pearl Lagoon, without having to go out to sea, just as one can go from Wankloa (north of Río Grande Bar) to Lamlaya (south of Puerto Cabezas), or from Sandy Bay to Wani Bar. In all, the inland lagoon–river system allows stretches totaling two-thirds of the coast from Bluefields to Cape Gracias a Dios to be traveled without going to sea. This is an important factor during the months of July, August, November, and December when high winds and rough seas make outside coastal traffic somewhat tense.

Coastal lagoon and estuarine biotic communities are composed of endemic species, and, depending on the nature of aquatic environmental conditions, marine species, and possibly a few osmoregulatory species with the capability to range to and from freshwater environments. The alternations between periods of high and low rainfall and different dominant winds, produce an everchanging diverse assemblage of aquatic species, capable of replacement and migration.

The lagoons experience a complex seasonal distribution of salinity levels which influences local fish and shrimp migrations and their avilability to the Miskito. Extensive flooding during the rainy season can make the coastal zone one vast sheet of water, flushing out all saline water from the lagoons and even drastically lowering salinity levels along the beach where the littoral current is fed by river outwash. During the dry season when the river levels drop, exposing numerous sand banks and swirling shoal waters, salinity levels increase as the tides are able to push in from the sea. Coincident with the changing salinity levels, water turbidity, and temperature, are radical shifts in aquatic life. During high salinity periods, large schools of shrimp and marine fish enter the lagoons, whereas freshwater fish with a low tolerance of salt water go up the rivers. When the rains come and the salinity level is lowered, freshwater fish once again come into the lagoons from the rivers. Thus, lagoon and estuary areas contain different fish at different times of the year and in different concentrations. During some months when conditions are favorable, there may be high populations of fish, while under extreme conditions there may be only a few species and in limited amounts.

The beach–lagoon–swamp environment is a zone of high biological productivity. The lagoon waters, as well as the river bar areas, are enriched by the discharge of nutrient-laden waters providing rich feeding grounds. Within the zone some of the most important fish utilized for food are snook (*Centropomus* sp.), catfish (*Ariidae*), mojarra (*Cichlasoma* sp.), guapote (*Cichlasoma* sp.), jacks (*Caranx*), and sheepshead (*Archosargus*); shrimp (*Penaeus* sp.), are also occasionally eaten as are oysters (*Crassostrea rhizophora*). In addition, the shallow lagoons and nearby river courses are favorite habitats for manatee (*Trichechus manatus*), which ranks as one of the most esteemed meats for the coastal Miskito.

Even though the water environment is dominant in this ecosystem, in some areas of the coast a variety of game animals is available in the beach forests, old and new plantations, and in palm swamps, including white-tailed deer, white-lipped peccary, paca (*Cuniculus paca*), and agouti (*Dasyprocta punctata*). From February through April freshwater hicatee turtles (*Pseudemys* sp.) and iguanas (*Iguana* sp.) are important in the Miskito's diet.

Coastal plant communities exhibit characteristic zonation found along many Caribbean beaches. Beach morning glory (*Ipomoea pes-caprae*) is

FORSYTH LIBRARY
FORT HAYS KANSAS STATE COLLEGE

a frequent pioneer in open areas above the high water level. Often a bush zone follows with sea grape (*Coccoloba uvifera*) and coco plum (*Chrysobalanus icaco*) being conspicuous. Next a narrow strip of coconut palms (*Cocos nucifera*) and then a closed herbaceous bush, shrub, and woodland zone are found as one goes inland (Fig. 19). Much of the former beach forest has been removed and replaced with secondary species because of extensive utilization of these areas for agriculture. When seen from the air, much of the beach vegetation appears striated because of the alternation of different species from beach ridge crest to trough. Beyond the beach area there commonly occurs either a palm swamp, marsh, or red mangrove swamp. These are frequently narrow belts of dense masses of vegetation. Swamps are made up of three types: herbaceous, forests, usually with palm associations dominating, and mangrove (*Rhizophora mangle, Conocarpus erecta, Laguncularia racemosa*).

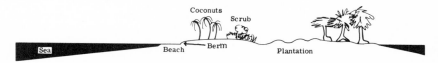

Fig. 19. Cross section of beach.

Subsistence agriculture takes place along the beach in the areas of beach ridges or any other places with even the slightest elevation. In these areas relief is small and only 1 yard may be the difference between beach ridge crests and troughs. According to Alexander (1965), who made a soils survey of northeastern Nicaragua including some of the beach soils, drainage is typically excessive, the soil is almost all sand, and nutrients are low.

Shallow Offshore Waters

Man's point of view is curiously different in the forest and in the sea. In the forest he is a bottom animal, in the sea a surface animal.
—Marston Bates, in THE FOREST AND THE SEA (p. 22)

Looking out of a boat on a sunny day on the groves of coral, sea-fans, sponges, and polypi, with the brilliant colours dancing in the unsteady water, and gaudy fish gliding about among the branches, one can imagine himself looking through some brilliant kaleidoscope.
—Charles Napier Bell (1862:268)

An extensive marine shelf extends far out into the Caribbean Sea off eastern Nicaragua. The shallow, warm tropical waters are strewn with hundreds of coral reefs and cays. The shelf is widest off Cape

Gracias a Dios, going out 75 or more miles and gradually decreasing in width southward toward San Juan del Norte. It has a very gentle gradient and rarely exceeds 50 fathoms in depth anywhere (Radley 1960:5–6).

This area has the largest sea turtle feeding grounds anywhere in the Western Hemisphere (Carr 1967:98), dominated by *Zostera* and *Thallasia* turtle grasses. These marine grasses are efficient producers of energy and make important contributions to the productivity of tropical waters (Odum 1971:357). The underwater marine pastures, or "turtle banks" support remnant populations of the once abundant Atlantic green turtle (*Chelonia mydas mydas*). The turtle banks may occur as isolated patches, 2 to 3 miles in diameter, or in large areas where the banks are close together. The vast expanses of marine grasses lie at a depth of about 10–20 fathoms; the air-breathing turtles dive to the bottom to crop the dense undersea mat and then rise 10–20 minutes later to "blow" (Fig. 20). Before nightfall, the turtles swim to nearby coral "shoals" to sleep.

Much of Coastal Miskito food procurement activity, such as patterns and amounts of labor inputs, timing of meat-getting pursuits, and distances traveled, are closely adjusted to green turtle migratory patterns and habitats.

The vast Miskito Banks lie off Nicaragua's northern Caribbean coast. This is one of two major turtle fishing areas along the coast; turtling activities are centered around the Miskito Cays, 30 miles east of the Sandy Bay villages (see Fig. 10, p. 50). Southward from the Miskito Cay–Miskito Bank area there are few cays or reefs, until Fox Reef off Wounta and Man O' War Cays 15 miles off Little Sandy Bay and Río

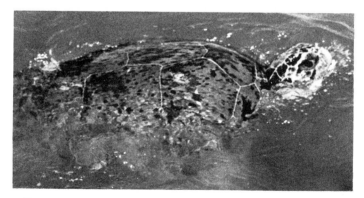

Fig. 20. A green turtle surprised while coming up to "blow"; turtle banks off Tasbapauni, May 1969.

Grande, where turtles are also taken. From the Man O' War Cays to the Pearl Cays (sometimes called the Set Net Cays) is the second largest expanse of cays, reefs, banks, and turtling grounds. Besides Man O' War Cays, other foci of turtling activities in this area include King's Cay, Tyara, and "Klar" (Crawl) Cays, 16–25 miles northeast of Tasbapauni; and Pearl Cays located 3–10 miles off Set Net. The Pearl Cays are numerous and close to the mainland, protected from strong swells and currents by Long Reef, lying to windward. This allows Set Net turtlemen to go out during rough weather.

South of Pearl Cays are only scattered cays, banks, and reefs, the most important of which are Pigeon and Frenchman's Cays off Honesound Bar at the southern end of Bluefields Lagoon, which is fished by the Rama Indians. A few other cays are found off Monkey Point, but they are only occasionally visited for bird eggs. A small turtle bank (Greytown Banks) located off San Juan del Norte is frequented by hawksbill turtles (*Eretmochelys imbricata imbricata*).

The Tasbapauni turtling grounds lie offshore, about 3–30 miles from the village (Fig. 21). The Miskito identify 20 "banks" ranging in size

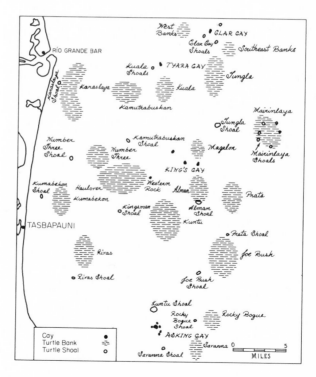

Fig. 21. Location of some of the most important turtle banks and turtle shoals recognized and exploited by the Tasbapauni Miskito.

from 5 by 3 miles in diameter to smaller banks, 1½–2 miles in width. These are scattered about within an area of some 600 square miles. The Miskito say that each turtle bank has one or more corresponding shoal areas, located from one-half to 3 miles from the individual banks. The complex of 50 or more shoal areas (turtle sleeping grounds) and 20 banks (feeding grounds), and several coral cays make up the locational features which determine much of the spatial pattern of the Miskito's exploitation of the sea. Because of distance, current, wind conditions, and luck, two turtlemen in a *dori* may stay out from 1 day to a week, each night sleeping on a nearby cay. During calm weather, when there are plenty of turtle, as in March and April, some turtlemen may stay on the banks and cays for weeks.

Starting in April and lasting until June, mature male and female green turtles migrate to Tortuguero (Turtle Bogue) in Costa Rica for mating and egg laying.[16] Tortuguero is the last remaining major nesting beach in the western Caribbean. The turtles return to the Nicaraguan turtle banks starting in late July and continuing through September. Thus a large percentage of the Miskito's food base is removed seasonally each year.

The marine fauna of the offshore waters correspond to the species found throughout Caribbean waters. The distribution of species varies with differences in bottom, depth, tubidity, salinity, and available food resources. There are few seasonal or annual changes in composition of most marine habitats, with the exception of the beach surf zone and adjacent littoral waters subject to the outpourings of fresh water from the rivers. Here salinity levels, turbidity, and currents are extremely variable throughout the year due to wave action and outwash of fresh water and sediments from rivers and lagoons. Open beaches offer little shelter from strong littoral currents and from predators. Fish and shrimp species come and go depending on local conditions.

There are three important fisheries in the area: shrimp, lobster, and turtle.[17] Shrimp and lobster are almost entirely exploited by commercial companies based in Bluefields, Puerto Cabezas, and Corn Island, and managed by foreigners and Ladino nationals. The turtle fishery, on the other hand, is the domain of the Miskito Indians and probably has been for much of their existence on the coast.

The hawksbill turtles are not sought after for food by the Miskito but for their shell which has a high market value. Hawksbill populations do not migrate to the extent that green turtles do, so there are many

16. Much of what is known about migrating patterns of green turtles is a result of Archie Carr's turtle tagging program at the Tortuguero nesting beach which has been in operation since 1956. See Carr (1967) for a description of what is known and not known about green turtle migration.

17. Craig (1966) discusses in detail the fisheries of British Honduras, many of which are similar to those found off the Miskito Coast.

egg laying sites on the cays and along beaches. The most significant hawksbill nesting beach in Nicaragua is on the mainland, on a deserted stretch of beach just north of Cocal, about 25 miles north of San Juan del Norte.

The shoal areas immediately adjacent to the cays are made up of coral sand and coral head bottoms. There are often large coral flats that may be uncovered during low tide. Shellfish (*Tegula, Strombus*), lobster (*Panulirus argus*), and fish are frequently taken from the shoals and flats for food by the Miskito during their turtling expeditions. The reefs are a very distinctive habitat and are often populated by fish found nowhere else. However, the Miskito seldom bother with the reef areas or the stretches of open water between the banks and cays, or between the beach and the banks, unless a turtle is seen there.

Except for the Corn Islands, there are no permanent settlements on any of the cays. During the height of the turtle season, temporary camps may be set up on some of the larger cays, however.

The cays are usually fringed with red mangrove or sea grape depending on exposure, and have coconut palms, *sani* (*Hibiscus tiliaceus*), coco plum, and buttonwood (*Conocarpus*) in the interiors (Fig. 22).

Fig. 22. A view of Asking Cay, southeast of Tasbapauni, 1971.

BIOTOPE AND SPECIES

The Miskito interact with biological communities in several different ecosystems, aquatic and terrestrial. The tropical rain forest, pine savanna, beach–lagoon–swamp complex, and the shallow offshore waters form the coastal environment of the Miskito. Within this overall area, each Miskito community exploits various biotic resources for subsistence found within the effective territory of the village. The population of a Miskito community, their territory, and the biological physical features found within, comprise a system in which material and energy exchanges are made between humans and their environment, and which form their ecosystem. However, human populations do not interact with their entire environments, or complete ecosystems, but rather with selected, well-defined components: biotopes (microenvironments) and certain biotic species. It is these components of a Miskito community's ecosystem which are most important for their subsistence and livelihood.

Within any human population's ecosystem, resources are unevenly perceived, unevenly known, unevenly distributed, unevenly available, and unevenly exploited (Dansereau 1966:448). Concentrations of sought after game animals, fish, or edible wild flora commonly occur in restricted areas, under specific ecological conditions and at certain times of the year. On the other hand, particular species of fauna may range widely between different biotopes and be the focus of exploitation rather than the biotope itself.

Biotopes are ecological units wherein primary habitat conditions, and fauna and flora adapted to them, are uniform. "Various habitats, or *biotopes*, can be distinguished, according to soil, vegetation, climatic conditions, each inhabited by a definite and well-characterized animal community [Allee and Schmidt 1951:4]." Each biotope is differentiated by distinctive physical features and biota such as a mangrove swamp, a *Thallasia* turtle bank, or a river sand bar where various reptiles lay their eggs.

Use of the concept of biotopes is important in this study because it provides definable segments of the environment which can be analyzed with respect to the subsistence system of the Miskito in terms of yields, species, and distances from the villages.

Research on other aboriginal groups in Central America has shown that Indian hunters have an acute perception of those segments within their environment which include concentrations of food resources. Bennett (1959) noted that the Chocó of Panama "did not range haphazardly through their habitat in search of fish or game but concentrated their efforts upon certain portions of the habitat (biotopes) which contain the desired animal species [p. 41]." Similarly, Coe and Flannery (1967), working on archaeological evidence on the Pacific lowland Maya in Guatemala, indicate that:

> Human communities do not react or interact with entire biomes. . . .The average lowland Maya, for instance, does not behave in relation to a broad and ill-defined "tropical forest" — he behaves in relation to small segments within it, and it is these that impinge on his life. . . .the stretch of forest where he hunts peccary and spotted cavy, the waterhole or stream in which the tapir wades: it is small microenvironments or *biotopes* such as these that determine the possibilities or limitations of culture change and population expansion [p. 7].

Since hunting and fishing are extremely important in Miskito subsistence, the distribution, accessibility, and dependability of animals is of tremendous concern. The Miskito recognize and exploit in varying degrees, the biotopes presented in Table 9. In Fig. 23 a diagram of the distribution of the biotopes is given, along with some characteristic fauna.

The number, diversity, location, and biotic richness of biotopes influence the time it takes the Miskito to procure fish and game, and wild

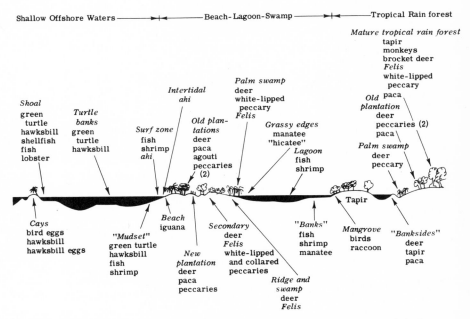

Fig. 23. Schematic cross section of major biotopes and associated characteristic fauna. Pine savanna biome not depicted.

TABLE 9

List of Major Biotopes Found on the Miskito Coast

Tropical rain forest and gallery forest

    Mature rain forest (*unta tara; unta disang*)
    Secondary (*unta sirpi*)
    Plantations—old, new
    Gallery forest (*upa twakni*)
    Banksides or "vega" (*albangkia*)—nautral levees
    Rivers, open runs (*auwala*)
    Breña (*apanis*)—dense bush, thicket
    Floodplains
    Oxbow lakes (*lakun* or *madakura*)

Pine Savanna (*auas twi*)
    Creeks
    Creek edges
    Wet low-lying savannas—herbaceous plants, papta palm
    High savannas—*Pinus, Curatella, Brysonima*

Beach–lagoon–swamp
    Intertidal zone
    Beach (*auhia*)
    Plantations (*insla*)—milpa or swidden
      old (*insla prata*): little tended, partially abandoned
      new (*insla disang*): currently worked
    Beach forest, predominantly secondary
    Swamp
      mangrove: various types
      palm: various types
      marsh (*busukra*): various types
    Lagoon (*lakun*)
      creek and river mouths
      shoals, banks—shallow mud or sand flats
      bars—where lagoon empties into sea
      sand bars (*auika*)
      grass flats (*kuswa apaka*)—along edges of lagoons, creeks, and rivers

Offshore shallow waters, reefs, cays
    Deep water (*unra; kabo tara*)
    Cays (*ki*)—small coral islands
    Shoals (*buhni*)—shallow coral flats
    Reefs
    Banks (same in Miskito)—turtle grass-covered feeding grounds
    Mudset—shallow, mud bottom waters near shore
    Close inshore waters
    Surf zone

plants, and to a large degree, their success in doing so. The positioning of biotopes with respect to one another can be envisioned as an environmental gradient (Porter 1965:411–413). Similar to contour lines, the closer biotopes are to each other in distance, the steeper is the environmental gradient; and in contrast, the farther away they are, the more level the gradient. Whether the biotopes have a steep or level gradient is significant in the ecology of Miskito subsistence in terms of the accessibility to diverse wild food resources. Because of seasonal flooding of extensive lowland areas along the coast, the migration of green turtles, and the change in fish fauna species, the environmental gradient can change radically throughout the year. Dansereau (1966) points out that

> the environmental gradients upon which species and communities are ordained either steepen or smoothen at various times and places, thereby reducing utterly or broadening greatly that part of the ecological spectrum which offers the best opportunity to organisms of adequate valence [p. 459].

A close spacing of biotopes containing biota which are culturally and physically available to the Miskito, should provide an ecological advantage by allowing utilization of multiple microenvironments, thus lowering subsistence risks.[18] Not only should subsistence risks be lowered but the time expended in travel to different biotopes in search of game should be less where biotopes are close together. Diverse biotopes with diverse species which experience seasonal changes in occurrence and distribution and an adequate area for agriculture offer wide spectrum opportunities for cultural adaptation and subsistence resource utilization which may not create severe ecological disruptions.

Several animal species important in Miskito subsistence may crosscut many biotopes in their occurrence. Green turtles, whitelipped peccary, and white-tailed deer yield substantial quantities of meat to coastal Miskito communities. Patterns of Miskito seasonal behavior have been adapted to a large extent to match the spatial and temporal behavior of these animal populations.

---

18. As defined by Porter (1965) subsistence risk "is a settlement negotiated between an environment and a technology [p. 412]." It involves uncertainty and loss of crops or hunting and fishing returns.

> Just how much risk an individual or a community can tolerate . . . is a problem that each culture must solve. A community has institutional and technical means of coping with risk. . . . Danger to the individual can be decreased by sharing out risks, through dispersal of fields, timing of harvests. . . . We may assume that in the degree to which the situation is tenuous, adjustment to risk is the essential element in the articulation of subsistence with environment [Porter 1965:412].

Similarly, subsistence risk can be lowered by the scheduling of human food-getting activities to coincide with the most favorable seasonal or environmental circumstances.

Abrupt ecological changes may occur at the contact zones between ecosystems or between biotopes. At these transitional interfaces, or ecotones, biological productivity at the primary and secondary levels tends to be great. Ecotones such as the river bank, lagoon edge, seashore, savanna edge, and rain forest border, offer maximum local and seasonal availability to diverse plant and animal species and to diverse biotopes (Harris 1972:184). An area with a large number of ecotones tends to provide a spectrum of wild biota and habitats from which subsistence patterns can be selected and scheduled.

The concentration on and exploitation of particular biotopes and species, the numerous ecotones, and the scheduling of subsistence activities to coincide with seasonal biotic and environmental fluctuations, encourages access to diverse resources and habitats in the land and water realm of the coastal Miskito.

CHAPTER VI

THE AVAILABILITY OF
MISKITO FOOD RESOURCES

The land of hereafter of the Miskito is far superior to the vale of tears in which we live; it is well stocked with all their favorite game, fish, and other food, as well as drink. Green turtles are plentiful and may easily be caught, and in the forests are large droves of peccaries and monkeys which may be killed at will. Fruit trees are continually in bearing and, like the other food plants there, do not need to be replanted. There are no other people there but Miskito; the paradise is unattainable to Sumu, Rama, Paya, and other tribes, or to foreigners.
—Eduard Conzemius (1932:154)

The Miskito inhabit a highly complex ecosystem. Food sources are abundant and varied yet the sum extent of wild and domesticated plants and animals includes many which are culturally unavailable for food, while others are accessible for limited periods only and, even then, in diverse places. A similar circumstance obtains with respect to the portion of foodstuffs purchased from the "outside," since some are not economically available to all, nor in equal amounts, nor consistently throughout the year.

The Tasbapauni Miskito interact in an intricate web of matter and energy exchanges with their environment and with plant and animal communities. Energy sources are not uniformly distributed in any ecosystem but occur in diverse structures which, in terms of human populations, are delimited by: (1) perception and desirability, (2) availability or seasonality, (3) dependability, and (4) productivity. The first two will be discussed in this chapter.

For any given population only a restricted portion of the energy in a given ecosystem is potentially utilizable or even needed. Most animal populations have energy consumption limits which are biologically imposed and relate to the composition and structure of food chains, population densities, or the occupancy of a behaviorally delimited niche. Human populations through cultural adaptation may increase territorial range, the kinds of materials that are utilized, or the efficiency and periodicity of extraction. Cultural considerations may encourage utilization of some resources while discouraging others (Weiss 1971).

For a subsistence people such as the Miskito, cultural desirability and temporal availability of specific foods are both major factors serving to integrate occurrence with scheduling of subsistence activities and with the ecology of the local ecosystem. Through cultural food choice and preference, the subsistence system is adjusted to conform with particular goal-satisfying food resource alternatives out of the wide-range of possibilities and temporal–spatial arrangements. The seasonal and areal availability of wild food resources and a culture's ability to modify or create habitats for domesticated species, constitute two of the most important environmental–cultural factors for cultural adaptation.

Food resources and their availability are a central point of reference in Miskito society and form part of a cultural ecological system influencing behavior. For individuals from industrialized societies it may be hard to imagine the intensity and primacy that food resources have in the everyday life of the Miskito. In her study of labor and dietary patterns of the Bemba of Northern Rhodesia, Richards (1961 ed) tried to emphasize this point:

> For those who are accustomed to buy food ready prepared, it is difficult to realize the emotional attitudes to foodstuffs among peoples who are directly dependent on their environment for their diet [p. 44].

> For us it requires a real effort of imagination to visualize a state of society in which food matters so much and from so many points of view [p. 46].

The Miskito "food year," with its times of hunger and times of plenty, is correlated with spatial–temporal changes in food availability. For the Miskito there are definite food seasons, comprising different resource complexes and strategies, labor inputs, locations, and degrees of subsistence risk. The seasonality of food resources and attendant labor patterns complement weather changes, thereby increasing the feeling of the passage of seasons.

CULTURAL AVAILABILITY

What Is Considered Food

After finishing a morning's work in a distant swidden, we were heating up some food for lunch, and one man told me: "Don't put dese beans in de record, dey is Spanish food."

The Miskito ethnoclassification of foods is an important factor in the ecology of their food system, for it not only delimits what is considered food, but it also classifies foods into a scale of preferences, which in turn determines the intensity and frequency with which specific wild plants and animals are sought after and which crops and crop combinations are cultivated. If a particular animal, such as the green turtle, is held in high esteem as food and is accessible, Miskito subsistence strategy will dictate that specific efforts be directed toward obtaining it. This strategy sets the pattern for labor arrangements, time devoted to turtling, and distance traveled to get the turtle; these, in turn, are all dependent in large part on turtle migrations, everyday whims of turtle behavior, and weather and sea conditions. Multiply this by the number of different food resources in the Miskito diet, each requiring specific strategies under varied ecological conditions, coupled with varying social responsibilities for food distribution, and one can begin to see the complexity of subsistence life.

The Miskito broadly classify resources into edible and inedible (Fig. 24). Only a restricted portion of the range of foods the Miskito have been exposed to through European culture contact has been incorporated into their diet. Rice and beans, a mainstay in the Nicaraguan diet, are considered "Spanish food" by the Miskito of Tasbapauni and are eaten only occasionally. On the Río Coco, however, rice and beans are much more prevalent in the diet due to greater contact with Nicaraguan Ladinos and widespread cultivation of beans. Wild meat—especially turtle, white-lipped peccary, and deer—is preferred over domesticated meat. All meat is classified in terms of whether it is fat (preferred) or lean, sweet or rank, clean or dirty, and whether it is from a "she" or "he" animal. Female green turtle meat is said to fatter, darker, and heavier than meat from male turtles, and consequently, eats better. Hawksbill turtle, collared peccary, and domesticated pig are considered to be rank meats, and therefore, less preferable. Green turtle and deer

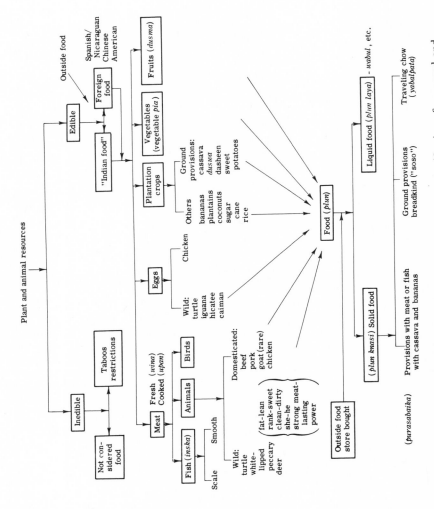

Fig. 24. Diagrammatic representation of Miskito ethnoclassification of prepared and unprepared foods, Tasbapauni.

106

are believed to have clean meat because they eat only clean plants. Shrimp are thought to eat mud, for that is where they live and nothing else is there; pigs and chickens are supposed to have dirty meats because of their nondiscriminating foraging habits around the village. Plantation crops are roughly divided into "ground provisions" (cassava, *duswa*,[1] dasheen) and others (bananas, plantains, coconuts). Some vegetables and fruits such as cassava, bananas, plantains, and mangoes, are often more desirable in their green or young state (*kura*), than when they "come to perfection" or mature (*tukla*). "Breadkind," probably a Creole word, is often used by the Miskito to include ground provisions, bananas, plantains, and breadfruit. Turtle meat is an item having a whole range of subtle preferences and categories built around it. Distinct sections of the animal are graded with respect to preference though everyone wants a piece of each type of meat.[2]

Turtle meat and cassava are the most highly regarded and sought after foods in Tasbapauni. Both are relatively dependable and high yielding; turtle is the major source of animal protein and cassava is the primary carbohydrate supplier. These two foods, one from the land and the other from the sea, form the core of subsistence in Tasbapauni.

A "perfect meal" for the coastal Miskito would consist first and foremost of meat, especially turtle, white-lipped peccary, or fish (roasted or browned in coconut oil), and boiled young cassava, green bananas, *duswa*, and some *wabul* (a thick porridge made from boiled green bananas mixed with coconut milk). Highly sweetened coffee and a bread made from flour would complete the meal if the individual was fortunate enough to be able to buy them. The Miskito say that cassava can be eaten easily in large quantities without becoming gummy and hard to swallow as are most other root and tuber plant foods. Other meats held in high regard by the Miskito are white-tailed deer, paca, agouti, manatee, hicatee freshwater turtle, and fish of many kinds, but principally, catfish, mojarra (*tuba*), stingray, and coppermouth. Cassava, green bananas, plantains, *duswa*, dasheen,[3] and coconuts form the vegetable core of the Tasbapauni Miskito diet and subsistence agriculture.[4]

1. Many varieties of this plant (*Xanthosoma* sp.) are widely cultivated along the Caribbean lowland for their tubers. It is also commonly referred to by its Jamaican Creole name "coco."

2. "Mixed" turtle meat consists of pieces of front or rear shoulder, calipash (the cartilaginous greenish substance that lines the inside of the upper shell), intestines, lungs, liver, and immature eggs if the turtle is a female and egg-bearing. Turtle meat is usually exchanged, sold, and prepared in mixed form.

3. Dasheen (*Colocasia esculenta*) is a fairly recent introduction to Tasbapauni. The first dasheen plants were said to have come from Orinoco, a Black Carib village located on the other side of Pearl Lagoon. Very few other Miskito villages have dasheen. It is still of only secondary importance in Tasbapauni, but it is increasing in occurrence due to its high yields and ability to grow in wet lands.

4. Tasbapauni food resources are listed in Appendix B.

The Significance of Meat

You got to have meat to eat. Wit out meat plenty people don't make fire morning to evening. Hungry. If you lucky and you get meat, all dem kitchen fires going wit dat meat. People happy.

—Tasbapauni, 1969

Meat occupies a prominent role in the everyday life of Tasbapauni.[5] In fact meat is the center of interest in the village and all other foods and activities are subordinate to meat getting. If there is one overriding concern in the life of a Miskito in Tasbapauni it is meat. As one Miskito put it: "Everyday you have to study down how you going to get de meat." Good times are described as when there is much meat in the house and in town (*upan kira*), and when there is no meat (*upan wauaia*), times are bad. One common phrase typifies much of the subsistence orientation of the Miskito of Tasbapauni to the sea and to meat: "*watla puban dingkisa wina*," or there is a high tide of meat in the house, which means literally that there is a lot of meat, and therefore food in general in the house.

Deprived of meat, many women refuse to cook or they are indifferent about it at best. In mid-July 1969, after many meatless days in Tasbapauni, everyone was watching for the return of some hunters who, hopefully, would bring back meat. One informant told me, "Dey want meat, no study tomorrow, dey say dat another day. Plenty not cooking. You no see smoke in dem kitchens dem. Some only have 'soso' [provisions without meat]."

Bell (1899), who traveled extensively among the Miskito for a number of years during the last century, observed that

> the Indian women estimate a man principally by his skill in hunting and fishing, while the agricultural department more properly pertains to the care of the women. It is a great reproach to a man if it is said of him that his wife suffers "meat hunger" (*opanwowaya*) [*upan wauaia*], an expression to indicate the craving which comes over one when long confined to a vegetable diet [p. 177].

With respect to the desire for meat, Bell (1899) went on to note:

> I could never imagine how the vegetarians of England get on, because when we had a week of plantains and maize and chili-pepper, often without salt, the craving

5. The Tasbapauni Miskito do not consider all animal flesh as belonging to the same generic classification "meat." A subtle distinction is made between animals which are regarded as having *real meat*, and lesser animals which have second-quality meat.

Meat is referred to in Miskito by three words: *wina*, flesh or uncooked meat; *upan*, butchered or cooked meat; and *tama*, meat which is put on the table as part of the "complete meal." Meat is considered so indispensable for a midday meal (the major meal) that when one says he has *upan* in the house he also means he has vegetable foodstuffs. If a Miskito says *tama apu*, or the meat is all gone, that means there is not food at all, whereas *upan apu* means there is no meat.

for fat or meat became unbearable; and when to that was added the taunts of the women and the pleadings of the children, no good man could stay at home, whatever might be the discomfort of the hunting path. I often pitied men who went out in the pouring rain and returned in the evening wet, shivering and hungry, and bringing perhaps a partridge, a tortoise or two the size of one's hand, and a couple of agouti, from which it was impossible to distribute a mouthful apiece among the villagers [p. 246].

Many investigators who have worked with traditional societies in many other areas of the world have remarked on their strong desire for meat and that it occupies a preeminent position in dietary preference. For example, Woodburn (1968) in describing the Hadza, a small group of nomadic hunters and gatherers in Tanzania, notes a similar longing for meat and its attendant importance in subsistence patterns:

> Although vegetable foods form the bulk of their diet, the Hadza attach very little value to them. They think of themselves and describe themselves as hunters. . . . In addition to being the preferred food, meat is also intimately connected with rituals to which the Hadza men attach great importance.
>
> Moreover the Hadza place such emphasis on meat as proper food and treat vegetable foods as so thoroughly unsatisfactory in comparison that they are apt to describe themselves as suffering from hunger when they have less meat than they would like [p. 52].

Fig. 25. Turtle butchering in Tasbapauni. Portions of the turtle meat will be set aside for the turtlemen's kin; the rest is sold for 50 centavos ($.07) per pound.

The significance of meat in Tasbapauni can be seen whenever an animal is butchered. Animals taken in hunting are brought to the hunter-man's house and cut up. When there are turtles they are butchered in the morning on the beach and a representative of each household is sent to get some turtle meat. Turtle butcherings are a happy event for the villagers and are one of the major occasions for group social exchange in the village (Fig. 25). Everyone gathers around the turtleman as he chops up the meat with a machete on a bloodstained board laid across his *dori*. No matter how many turtle butcherings the villagers have seen before, they will be there, to comment on whether the meat is fat or lean, if it is a she turtle, and if it has eggs or not.

Food Taboos, Restrictions, and Beliefs

Within the range of available foods there are some that cannot be eaten by certain people because of taboos and restrictions. For example, many Miskito have a charm or "body guard" called *kangbaiya*[6] which protects them against spirits or the bad intentions of others. In order to insure that a *kangbaiya* remains effective, the holder must, among other things, observe certain food prohibitions (Table 10). Not everyone has a *kangbaiya*, nor do all rigidly observe the food prohibitions; however, compliance is sufficient to influence significantly the direction of some hunting and fishing efforts and dietary patterns. There are also short-term food avoidances which pertain generally to pregnant women. Similarly, one who has been treated by a *yumu waikna* (a lesser shaman-type specializing in stomach aches) will be told to stay away from certain foods. For example, some of the foods one should not eat after being cured of a stomach ache (*yumu*) are oils, sugar, salt, catfish, and beef. Usually the prohibition lasts for only a few days. Individuals will not eat certain foods which will go "slingwi" or against a particular bush medicine. Foods which will defeat the properties of the medicine are many, but limes, chili peppers, and garlic are especially singled out.

It may appear logical to expect a society to utilize an environment to its entirety in terms of available technology but it is not done. No society fully exploits the possible food resources of its environment to the extent its technology would permit. Sanctioned behavior and breadth of perception reduce the utilization of natural resources to a small number of possible choices. Traditional Miskito perception and prefer-ence of food resources increase extractive efficiency, and encourage a relatively well-balanced diet. There are no restrictions against green

6. Called *kangbaika* on the Río Coco.

TABLE 10

Food Avoidances. Some of the Not Eaten, Rarely Eaten, or Restricted Plant and Animal Foods in Tasbapauni

Not food or do not eat	Rarely eaten	Not eaten during Pregnancy	Kangbaiya
Opossum	Squash, pumpkins	Young coconuts	Hawksbill turtle
Howler monkey	Tomatoes	Pineapples	Collared peccary
Anteater (2)	Beef	Melons	Lobster
Squirrel	Brush rabbit	Sugarcane	Shrimp
Sloths (3)	Pork	Ripe fruits, i.e., banana	Jack fish
Cats (5)	Lime	Plantain or banana	Tarpon
River otter	Soup	porridge—"wabul"	Mullet
Bush Dog	Milk	Green turtle head	Chicken
Snakes—various	Chicken eggs	Hicatee turtle	Tapir
genera	Shrimp		Coati-mundi
Many fish	Vegetables	*After delivery*	Racoon
Islu lizard	Corn		Curassow
Salati crabs	Sweet potato	Breadfruit	Armadillo
Frogs and toads	Parrots, macaws	Beans	
Trunkback and		Coconut	
· loggerhead turtles		Shrimp	
Porpoise			
Jewfish			
Weasles			
Kinkajou			
Pelican			
Caiman, crocodile			
Leafy greens			

turtle or white-lipped peccary; all fairly large animals with high density populations. Miskito food preferences and taboos for wild fauna are generally against small animals, having low densities, thereby serving to channel meat-getting efforts toward high return animals. As we shall see later, Miskito hunting and fishing efforts are very efficient because they concentrate on large animals in localized areas and on small animals having high concentrations. Under low human population numbers and densities and with faunal exploitation for subsistence alone, Miskito hunting and fishing foci were geared to high dependable returns without causing massive ecological disruption (Nietschmann 1972). Wild animal resources have an ordered desirability for the Miskito and receive a concomitant degree of hunting and fishing focus. To some extent food preferences may be thought of as adaptive mechanisms for maintaining acceptable levels of exploitation. Food restrictions against economic animals such as pigs and chickens are also functional, assuring that these animals will be available for market.

White-tailed deer used to be a taboo meat according to Miskito informants, coinciding with an apparent widespread prohibition by many lowland tropical groups against deer meat.[7] Tasbapauni Miskito say that the acceptance of white-tailed deer meat has occurred only in the last 40 years or so. This date coincides approximately with the period of the exodus of men from the village to work for banana and lumber companies. With most of the men away, fewer turtle were brought into Tasbapauni to support the remaining villagers. It may be that meat-getting attention was turned to the white-tailed deer populations which were easily accessible in nearby plantations.

In addition to white-tailed deer, a few other taboo or low preference animals have recently been classed as edible, even desired meats by the Tasbapauni Miskito. For example, tapir (*Tapirella* sp.), manatee (*Trichechus manatus*), shrimp (*Penaeus* sp.), once undesirable food animals, are now eaten; manatee being one of the most sought after. This change may have come about as a response to the decline in other meat sources.

In the world of the Miskito there are a number of spirits, beliefs, and cultural constraints controlling or limiting the daily availability of many food resources. Many spirits and supernatural animals, said to live in the forest, rivers, lagoons, and at sea, act as a deterrent to hunting at night. This is when some animals are more easily taken, paca and deer for example. A few animals are supposed to have "keepers" who protect them and lead them. The *wari dawan* (white-lipped peccary owner or keeper), the *sula dawan* (keeper of white-tailed deer) will guard their animals from hunters who kill too many. According to the Miskito, these owners have certain birds which act as sentries and if a hunter hears the call of that particular bird he should turn back, for if he continues the *dawan* will lead him astray in the forest. Bad spirits such as *patas, waiwan, prahaku, wakumbai, liwa, li lamia, lasa,* and *waula tara*, are all considered dangerous but do not have much influence on hunting or fishing unless an unexplainable incident occurs; then people will stay away from that particular area.[8]

Some of the younger men no longer believe completely in the existence

---

7. For example, Bennett (1962) reported that according to Wafer the Bayano Cuna of Panama "never killed deer nor ate deer flesh, and gave evidence of displeasure when some of Wafer's party killed and ate a deer. . . . The deermeat taboo was still in effect as recently as the mid-1930s [p. 41]." Similarly, Carneiro (1957) noted that the Kuikuru of Brazil considered deer meat a tabooed food item.

8. *Patas*, a spirit with light that lives in the forest; *waiwan*, an invisible evil spirit animal that lives in all places—forest, sea, lagoon, river; *prahaku*, similar to a mermaid and able to tip over dories; *wakumbai*, a spirit rider on a mule-like animal, very dangerous; *liwa*, water nymph; *li lamia*, water tiger; *lasa*, a general word for ghosts of many kinds; and *waula tara*, a large boa which is said to live at the headwaters of streams and able to make the level fall by sucking up water.

of spirits. However, if a hunter gets lost, or is cut up by a drove of white-lipped peccary, then everyone attributes it to a spirit or keeper and everyone believes again.

When a person dies it is considered a sign of respect and community concern to stay home and not go to the plantation to work, nor go hunting and fishing. When there is a "dead" it is a time of trouble and people try to help. Men build a wooden casket, women cook food and console the family. Thus with the high death rates in most communities, many days of potential food-getting activity are removed. Similarly, no food is sought from outside the village on Sundays, Easter Week, and Christmas Week; people are supposed to stay home and visit, or go to church.

The tide and phases of the moon influence the timing of subsistence behavior. Certain times of the month and day are preferred for hunting, fishing, and harvest. The Miskito feel that "all animals work with the tide." The best times to go hunting or fishing is on a rising tide when animals are active, feeding, moving about. Most animals are believed to go away to sleep when the tide is falling. Hunting and fishing activities are planned to coincide with a favorable tide. This in effect reduces the amount of time available to hunt or fish, but in the eyes of the Miskito it may increase the efficiency of hunting and fishing through a higher rate of return for hours invested. According to the Miskito, there are many indicators in the forest to show what the tide is doing. One of the most important are tide birds, especially the woodpecker who is said to peck when the tide is changing.

Another influential element in Miskito subsistence efforts and the availability of foodstuffs is the moon. For the Miskito "the moon controls all." Different stages and changes of the moon are closely watched. In general, a new moon (*kati apu*) is considered bad for all subsistence activities. White-lipped peccary are said to walk all night; manatee are very restless and are difficult to find; deer are believed to be very nervous to a hunter's approach; most crops will spoil or be light in weight if harvested during this moon phase. From the first quarter to the last quarter is a good time to go hunting, fishing, or to harvest. The Miskito say it is best to go after most animals and crops during such specific moon phases.

The moon and the tide act as general indicators for the best hunting and fishing times, and to a lesser degree, for agricultural pursuits. They are not rigid constraints and many Miskito will go "against them" if need for food is great. For the most part the Miskito's recognition of the coincidence of animal behavior and the tide (time of day) or moon stages restricts their food-getting activities to what is believed to be the most productive periods.

Additional Factors

"Meat in the market, ten cents a pound; broke foot in Cabezas, can't get a pound."
—Song once sung by Miskito
wage laborers (circa 1920–1930)

A portion of the Miskito's food comes from outside the traditional subsistence system economy and is purchased from small "stores" in the village, or from fellow villagers. Consequently, various store foods and local foods are available to those who can afford to buy. As most money is obtained seasonally from hunting (skins, meat), fishing (hawksbill shell, shrimp, green turtle, calipee), and agriculture (rice, coconuts, coconut oil), the volume and intensity of purchases varies accordingly. Most families can get together 50 centavos to 1 *córdoba*[9] a day for flour, sugar, coffee, but some cannot and do without.

A number of other factors influence or determine the availability of foodstuffs to particular individuals in Tasbapauni. Kinship ties and obligations particularly affect food getting, especially if one is close kin to a few successful turtlemen. A family with a small range of kinship ties has a reduced potential for acquiring food.[10] A family without a *dori* is handicapped in food-getting activities as many food resources and agricultural areas can be reached by water only. Often arrangements are made for the loan of a *dori*, sometimes including the payment of a *"dori* share" to the owner, or one-fourth of the catch obtained in hunting or fishing. The absence of means of food storage beyond smoking or sun-drying of small quantities of fish and meat, limits the availability of many foods to their seasonal occurrence. Furthermore, a village's location with respect to a market town (Bluefields or Puerto Cabezas) determines somewhat the availability of store bought foodstuffs or a ready market for skins, coconut oil, rice, and shrimp. Isolated villagers, such as the Miskito who live above the rapids on the Río Coco, either buy store goods at greatly increased prices from local Ladino merchants or wait to make a 50-mile trip down river to San Carlos or 150 miles to Waspam.

Since 1965 the Nicaraguan government has attempted to enforce a law against taking turtles from May 15 through July 15, in order to protect migratory males and egg-bearing females. As few Miskito villages have a resident government *guardia,* the prohibition is ineffective. Turtles cannot be sold in Puerto Cabezas or Bluefields during this period.

---

9. One *córdoba* (C$1.00) equals $.14, 1970.
10. See Chapter VIII.

SEASONAL AND SPATIAL AVAILABILITY

In de dry de men tend to turtle and shrimps. In de rain dey go after de animals in de bush.
—Tasbapauni, 1971

Within the range of culturally acceptable and utilized foods, there are temporal and spatial differences in their occurrence. The food resource base is not a static, unvarying entity, available at the same time and place day after day. The seasonality and variability of food location presents a kaleidoscope of changes throughout the year. The Miskito can be thought of as operating within a dynamic spatial–temporal zone rather than just lineally ranging through their environment. Both wild and domesticated plant and animal resources mirror changes in environmental processes in terms of seasonality, location, and degree of accessibility.

Seasonality

Many of the subsistence foods of the Miskito are available at only fixed and predictable times of the year, or possess some special attribute making them more desirable than at other times of the year (Table 11). For a Miskito everything has its "time" or "season." Game animals, fish, wild fruits, plantation crops, practically all have a "time." As Conzemius (1932) observed, "the behavior of the elements, the opening of certain flowers, the maturing of certain fruits, the song of certain birds, and the spawning time of certain animals, each in its season, are thus the almanac of the Indians [p. 110]." And "the approach of the rainy season, or winter (*li mani*), or the dry season, is known from the behavior of the animal and vegetable world [p. 109]."

Many months of the year are referred to in Miskito by the name of the animal especially abundant and obtainable at that time. For example, February is called *kuswa kati* or hicatee turtle month; March is *kakamuk kati*, iguana month; April is *wli waikna kati*, male green turtle month; and May is called *wli mairin kati* or female green turtle month.

Some animals occurring more or less on a year-round basis are thought to be more desirable at specific times of the year because they are fatter (e.g., white-lipped peccary) or because they are with eggs (iguana, hicatee, green turtle).

Within the span of environmental and biological processes affecting the availability of food resources, flooding, salinity level changes, unfavorable winds and currents, and turtle migration can be singled out as being foremost (Table 12). Turtling and fishing provide the major sources of animal protein, and are a main focus of labor effort for the coastal Miskito. The desired species often have local or regional

TABLE 11

Seasonal Changes in Food Supply, Tasbapauni[a]

	Jan.	Feb.	March	April	May	June	July	Aug.	Sept.	Oct.	Nov.	Dec.
Cassava												
Coconut												
Rice												
Duswa												
Banana—plantain												
Breadfruit												
Mango												
Pineapple												
Pejibaye												
Green turtle												
Fresh water fish												
Salt water fish												
Shrimp												
Manatee												
Peccary												
Deer												
Hicatee turtle												
Iguana												
Ahi												
Ibo nuts												
Coco plum												
Sea grape												
Hicatee eggs												
Iguana eggs												

[a] Line symbols indicate the period each food resource is used or is available either because it matures during that specific month, can be caught then, or because only enough crops were planted to last for limited months. Only some of the important food resources are shown (supplies scarce – – – –; supplies available –––––––; supplies abundant ————).

116

TABLE 12

Some Environmental and Biological Processes Influencing the Availability and Accessibility of Important Food Resources, Tashapaumi

Process	Hawksbill turtle	Reptile eggs	Ahi	Iguana	Hicatee	Manatee	White-tailed deer	Peccary (white-lipped)	Green turtle	Shrimp	Fresh water and salt water fish, shrimp	Domesticated crops	Wild plants
Fruiting season	x	x										x	x
Nesting season												x	
Changing salinity levels				x					x		x	x	
Water temperature, turbidity										x	x		
Lack of available food resources										x	x		
Extensive flooding						x	x	x			?		
Unfavorable winds, currents	x		x		x	·x		x			x		
Local migration, movements	x				x	x	x	x	x	x	x	x	
Long-distance migration	x					x		x	x	x	x	x	
Lack of water during dry season									x	x			
Ecological conditions favorable—dry season	x	x		x	x	x	x	x	x	x	x	x	
Ecological conditions favorable—wet season								?				?	x

migratory habits, corresponding with salt–fresh water changes for fish, and local and long distance migration of green turtles. Similarly, other mainstays of livelihood occur seasonally such as rice and shrimp which are sold to outside markets.

Equally as important as the seasonality of food resources is whether or not they are "in phase" with work demands in other spheres of the subsistence system. Thus, the February through May dry season, when green turtles are the most plentiful and easily caught, is the period when agricultural work (clearing, burning, planting) has to be done. Energy-providing meat is generally widely available during this crucial time. The Miskito say that they need meat to make them strong enough to do hard agricultural work. The sea resource, turtle meat, supports in large measure preparations for food getting on land during the dry season. Changes in local ecosystem dynamics (seasonality, migration) directly affect the timing of a great deal of subsistence activities. The "fit" of these activities and cycles in the ecosystem will be discussed later.

Spatial Variability and Relative Location

The food resources of Tasbapauni vary not only temporally, but spatially as well. The availability and accessibility of foodstuffs are influenced in part by their relative location, distance from settlement site, range, and local distribution. Whether one perceived resource is used or not is often a matter of subsistence need, available market, and "variable cost" in terms of labor, time, distance, and degree of success, rather than polarized categories of "possible" or "impossible" behavior. Thus, the location of food resources in relation to the settlement site and to each other helps determine the degree to which they are utilized.

The location of agricultural, hunting, and fishing areas used by the villagers of Tasbapauni is shown in Fig. 26. The land–water zone within the range of exploitation totals approximately 850 square miles, which include 625 square miles of ocean, 125 square miles of lagoons, rivers, and creeks, and 100 square miles of land. Not all of this area receives the same intensity of exploitation, of course. The 850 square miles include all of the territory which is visited or traversed during the year for food getting.

The village has more land available to it than 100 square miles. In 1917 a representative of the British government visited many Miskito villages in order to legalize boundary claims. Ownership documents were also drawn up for some of the offshore cays (Asking Cay, Buttonwood Cay, Swiri Cay, Tangweras Cay, King's Cay, Kaymer Cay and Savanna

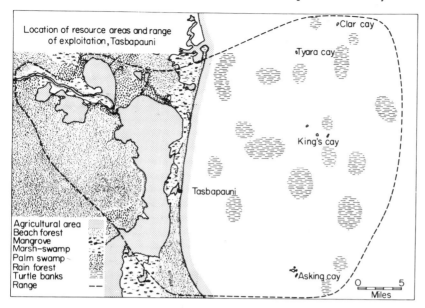

Location of resource areas and range of exploitation, Tasbapauni

Clar cay
Tyara cay
King's cay
Tasbapauni
Asking cay

Agricultural area
Beach forest
Mangrove
Marsh–swamp
Palm swamp
Rain forest
Turtle banks
Range

0   5
Miles

Fig. 26. Location of subsistence resource areas, Tasbapauni.

Cay).[11] The original land boundaries of Tasbapauni have been altered somewhat over the years through the disputed "sale" of portions of land to outsiders by "representatives" of the village.

The Tasbapauni people carry out almost all of their food procurement activities within the land area held by the village and off the village's cays and intervening waters. Neighboring villagers catch shrimp every year during February through June within Tasbapauni territory.

In terms of the absolute area exploited by the Miskito, water is obviously the dominant resource realm. The total for all ocean, lagoon, river, and creek areas which are exploited or crossed to get to other food resource areas, amounts to 85% of the 850 square miles (73% ocean; 12% lagoon, river, creeks) which comprise Tasbapauni's subsistence exploitation area. As the majority of meat supplies come from water-based environments, it is evident that distance traveled should be equated with *dori* travel.

Within the large land–water zone exploited by the Miskito of Tasbapauni, the availability and relative location of food resources differ most markedly between the dry season and the wet season. The loci of hunting, fishing, agricultural, and gathering activities are radically shifted during these seasonal extremes (Figs. 27 and 28). During the February through

11. Libro de Propiedades del Departamento de Zelaya, Palacio Municipal, Bluefields.

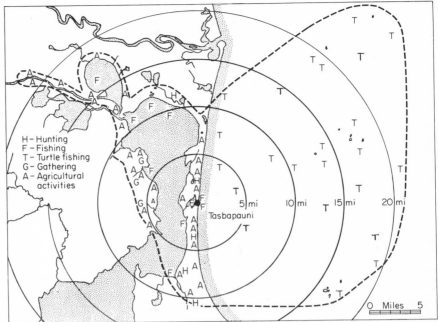

Fig. 27. Location of exploited food resources and associated activities during April (dry season), Tasbapauni.

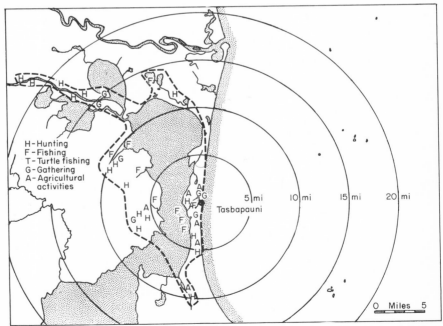

Fig. 28. Location of exploited food resources and associated activities during July (wet season), Tasbapauni.

May dry season, food resources are generally more diversified than in the wet season but agricultural foodstuffs are depleted. The food resources are also more widespread. Conversely, in the rainy season, resources are more limited in quantity and kinds, but the distance traveled to get to them is reduced due to limitations caused by flooding, bad weather, and animal migration. In October through December hunting activities range farther afield in search of game. This is the time of variable and strong winds, which seriously limit turtle fishing. Instead, increased emphasis is placed on white-lipped peccary and white-tailed deer whenever the men can get away from harvesting rice.

The distribution of animals varies between wet and dry seasons. As noted previously, at the start of the rainy season in mid-May, ocean and brackish-water fish move out of the lagoons and away from the longshore littoral current as river and creek runoff invades the coastal area. Freshwater fish reenter the lagoons from their river and creek dry season retreats (Table 13). During the wet season turtling activities drop off due to rough weather and the migration of most of the mature turtles. Hunting becomes the major meat-getting activity; fishing is difficult in the murky, flooded lagoon. Gathering assumes more of an importance than in other times of the year as many wild fruits ripen, while other food resources are at their lowest point of the year. The

TABLE 13

Differences in Some of the Types of Fish Caught during Wet and Dry Seasons, Tasbapauni

Area found	Dry season fish	Wet season fish
Lagoons	Coppermouth Snook Jack Drummer Sawfish Shark Shrimp	Catfish Stone bass Sheepshead Mullet, Califavor
Rivers-creeks	Catfish Snook Mojarra Guapote	Mojarra—few Guapote—few *Asamina*
Ocean	Jack Kingfish Barracuda Yellowtail Tarpon Rook rook Drummer Stingray Snook	Snook (almost no wet season ocean fishing done)

dry season is the time for turtle, shrimp, fish, and agricultural work in the plantations. Long trips are made after turtle and to riverside plantations 10–25 miles distant. Temporary camps are set up on the cays and up the rivers.

The Miskito have many emergency sources of food which they can fall back on when food is scarce because of bad weather, flooding, or whatever. During the height of the rainy season in July when it may even be impossible to walk up the beach to one's plantation to dig out a few roots of cassava, food shortages are severe. In Tasbapauni the most used emergency foods are *ahi* shellfish (*Donax*), dried and salted meat and fish, and hone palm seeds. What usually happens in Tasbapauni is that if some foods become unavailable, more pressure is put on less used food resources. Emergency foods are more important for the Río Coco Miskito who do not have the food resource range existing in Tasbapauni. If a rice crop or bean crop fails due to disease or insects, conditions can verge on famine. Starch from banana stalks, corn cobs, tree bark, the hearts of various palms, are all eaten.

Generally, then, during the dry season much of the food-getting activities are focused on marine resources and during the wet season on terrestrial resources. The location of Tasbapauni with respect to its land–water connectiveness and accessibility to diversified food resources in different ecosystems allows an almost unrestricted shift between resources and resource spheres; thus, there is a tendency toward a balancing of shortages and surpluses. The range and density of biotopes, the changes in their faunal and floral availability, and the various fruiting seasons and crop maturity periods (many with overlapping phases) permit seasonal changes in the exploitation pattern of the villagers, permitting as we shall see, a qualitatively and quantitatively balanced dietary pattern under subsistence conditions. The scattered biotope location of resources within respective environments necessitates a radiating exploitation effort with the village at the center. The land- and water-based resource inventory and the large area exploited require a high degree of mobility.

The distance between the settlement site and the location of the desired food resource represents an obstacle to interaction. The farther the resource, the more time it takes to get there, and generally the greater must be the assurance for success. The extreme distances the people of Tasbapauni travel to their farms (walking up to 6 miles one way; by *dori* up to 25 miles) and to hunting and turtle fishing areas involves an inordinate amount of inputs of labor and time. In general, with respect to distances traveled to farm sites, Chisholm (1962) stated that any distance up to about a kilometre from the dwelling is of such little moment for any but specialized farming . . . that little adjustment is called for in either pattern of settlement or land use. Beyond about 1 kilometre, the costs of movement become sufficiently great to warrant some kind of response; at a distance of 3–4 kilometres the costs of cultivation necessitate a radical modification of the system of cultivation or settlement . . . though adjustments are apparent before this point is reached [p. 148].

Thus we may expect to see a modification of food procurement systems accommodating factor of distance inputs. This will be examined in the next chapter.

A better idea of the temporal and spatial dynamics of the availability of food resources can be gained by considering in detail the locational characteristics of the green turtle, the most sought after food resource (Table 14). Green turtles are particularly remarkable animals on which to focus meat-getting actions. A mature turtle may weigh 250 pounds or more, graze untended on extensive sea pastures, have dependable habits, and when caught can live for months in a crawl, or for a few weeks on its back under shade.

Green turtles have at least six behavioral characteristics that are important for their exploitation by human populations: (1) they occur in large numbers in localized areas; (2) they are air breathing, so they have to surface; (3) they are mass social nesters; (4) they have an acute location-finding ability; (5) they migrate long distances; and (6) they display predictable changes in their spatial occurrence and distribution according to environmental conditions and to daily, seasonal, and annual cycles. It is these latter two points that we wish to discuss here.

The turtles move from the shoal areas off the cays to different "banks" or feeding grounds from month to month and during the same day. They change their distributional patterns depending on the time of year, sea and wind conditions, and even the time of day. The Miskito have to learn these habits in order to anticipate where they will find a congregation of turtles, rather than simply ranging back and forth at sea, haphazardly searching for turtles.

The turtles can be exploited along several paths or at several points in their patterns of movement: (1) seasonally, at different areas depending on weather and current conditions, (2) daily, at or between their sleeping and feeding areas, (3) leaving for or returning from long distance migrations, and (4) at their distant nesting beach.

The Tasbapauni Miskito distinguish five primary areas or zones where turtles congregate: (1) shoals or sleeping grounds;[12] (2) banks, the

12. Cayman Islanders refer to the sleeping shoals as "bars."

TABLE 14 *Monthly Turtling Conditions, Tasbapauni*

	Jan.	Feb.	March	April	May	June	July	Aug.	Sept.	Oct.	Nov.	Dec.
Location of turtles												
Banks near cays	X	X	X							X	X	X
Banks between cays and mainland		X	X	X	X	X	X	X	X			
Inside "mudset" zone				X	X							
Weather												
Wind strong	X	X				X	X	X	X	X	X	X
Wind light			X	X	X					X		
Wind direction	NE	ENE	NE / E	E / SE	E	E / SE	NE / E	N[a]	NE[b]	[c]	N / NE	N / NE
Sea rough	X	X	X			X	X	X	X	X	X	X
Sea calm	X		X	X	X							
Turtle fishing												
Strike	X	X	X	X	X	X		X	X	X	X	X
Nets		X	X	X	X			X	X			
Good fishing	X	X	X	X	X			X	X			
Bad fishing	X					X	X			X	X	X
Turtles taken[d] Oct. 1968–Sept. 1969	84	95	133	109	67	49	0	103	90	41	19	29
Condition of turtle												
Fat				X	X			X	X			
Lean						X	X					
Migration												
Leave				X	X	X		X	X			
Come back					X	X	X					

turtle feeding grounds between the offshore coral cays and the mainland, (3) "cays," distant turtle feeding banks near coral cays; (4) "mudset," close-to-shore mud bottom zones where turtles are found during the dry season; and (5) the distant Tortuguero, Costa Rica nesting beach.

The annual cycle of green turtle movement, although complex, is relatively foreseeable to a skilled turtleman. During January when the north and northeast winds are stiff, green turtles usually gather on the shoals and banks in the lee of the cays. As the dry season approaches and the winds slack off, and the freshwater littoral current abates, the turtles move from the area of the distant cays and spend more time on the closer banks, from which one can see the mainland. By April and May the turtles have moved in very close to shore, in the mudset zone, and many turtlemen are setting turtle nets 2 or 3 miles off the beach in addition to making long distance journeys to other turtling areas far out at sea. During the dry season many turtlemen will stay at temporary camps set up on some of the cays until they have accumulated a number of turtles.

Starting in early April the first group of turtles is said to leave for Tortuguero in Costa Rica for nesting. Turtles continue to leave during May through early June. Not all turtles depart for the nesting beach, however. Nesting takes place generally on a 2- or 3-year cycle. When the turtles do move off for Tortuguero, probably only the mature "she" turtles destined to lay for that year, and the "big, big he's," go. The Miskito claim that the turtles "move off in groups of he's and she's" and follow a definite "route" to Tortuguero: "Turtle navigate, you know; leave off at night and go direct for Turtle Bogue," as one turtleman remarked.[13] From June to the first part of August turtling activities diminish because of the absence of large numbers of turtles and because of adverse weather conditions. The remaining turtles stay out on the distant banks and shoals away from the up to 10-mile wide belt of freshwater river discharge pouring out of the many river and creek bars. The turtles begin to come back from nesting in late July and continue arriving until most are back by September. These turtle are said to be "meager"—without much fat. Returning females have fresh scratches on their top shell from a male turtle's frantic coupling efforts during mating and scratches on their bottom shell from crawling over debris on the beach to reach a suitable area above the high tide zone where they nested.

September with calm weather is a good turtle month; sometimes turtles are harpooned at night if the sea is flat and there is little wind. From

13. The story of how green turtles were shown to be one of the world's most amazing migratory animals is beautifully told by Archie Carr in his book *So Excellent a Fishe* (1967).

October through December the turtles stay for the most part on the banks near the cays sheltered from the seas running before the strong north–northeast winds.

In addition, wind conditions change the local turtle distribution patterns. Winds blowing from the northwest to northeast sectors cause the turtles to bunch up, increasing the turtlemen's chances of success. When winds come from the east to the southwest the turtles scatter. Infrequent west winds bring the turtles closer to land.

Green turtles follow what one might call "commuter behavior," a daily cycle of journeying from their sleeping shoals to their feeding grounds. At about 5:00 A.M., the turtles gather to leave the sleeping shoal where they have spent the night, for the banks, possibly 1–3 or 4 miles off. By 7:00 A.M. they can be found on the banks where they go through another cycle of float–dive–feed–surface–float, taking 15–30 minutes to complete the process. By 4:00 or 5:00 P.M. they begin to move off the feeding bank and return to their favorite shoal or shelter under a coral head or reef. The commuter behavior of green turtles is well-known to the Miskito and they will try to deploy themselves between the shoals and banks at 6:00 A.M. to intercept the turtles. Turtle nets are set over the sleeping shoals to entangle homecoming turtles or ones surfacing during the night to "blow." Similarly, turtlemen time the duration of feeding dives and floating intervals and attempt to reach the spot where a turtle is expected to surface, or reach a turtle before it dives again. The Miskito claim that turtles usually blow three times while floating on the surface. This seems to be true from my own observations. After swimming to the surface, the turtles blow air in a loud hiss which can be heard 40 or 50 yards away. They will dive after the third blow for air, which gives the turtlemen a minute or so to locate and reach the turtle after hearing the first blow.

Their acute knowledge of green turtle behavior assures the Miskito of advantageous turtling conditions, increasing chances for high returns, allowing for exploitation activities in other resource spheres during poor turtling periods. The location of turtles varies considerably throughout the year, but the patterns can be predicted by the Miskito, and their efforts directed correspondingly. Turtle fishing requires no cooperation among large groups of individuals for success. Turtlemen work as partners in turtle fishing, meat butchering, sharing, and sale. Turtle fishing is better during the long and short dry seasons (February through May, September) than at any other time of the year. Turtles are plentiful and weather conditions optimal. "Fair weather" turtlemen are able to obtain meat fairly easily, supplying much needed animal protein to large segments of the village during times of low agricultural output, and after three low meat months during the June through August rainy season. At other times of the year "real turtlemen" are the meat getters; turtling is harder and sea conditions rougher.

Turtle migration periodically removes a large part of the population away from Miskito exploitation pressure; unfortunately, though, localized nesting behavior concentrates adult turtles and new generations of turtles in situations where they are extremely susceptible to intensive commercial exploitation. This has happened for years at the Tortuguero, Costa Rica nesting beach (Carr 1956;1967).

Due to their predictable behavior and the many points in their patterns of movement where they can be caught, their availability has decreased because of long-term overexploitation for commercial concerns. Thus, even though the Miskito have adapted much of the scheduling of their subsistence activities to coincide with favorable turtling conditions, the drastic depletion of green turtles has made them less available, more inaccessible, and more of a subsistence risk. The decrease in turtle numbers and availability has brought about several responses from the Miskito, as they attempt to maintain acceptable catches of turtles for subsistence as well as increasing exploitation for market sale: (1) turtle dories are being made smaller and faster so that they can approach the increasingly wary turtles; (2) turtlemen are changing from harpoons to more efficient, but more costly turtle nets; (3) amounts of search-time for turtles is being increased; (4) distances traveled for turtles are increasing; and (5) alterations are being made in subsistence activities on land to permit these greater expenditures of time, distance, and money.

It remains here to emphasize that much more is encompassed in resource availability than simply perceived and ordered desirability of resources, and spatial and temporal environmental changes, affecting faunal occurrence. Skill, experience, and technological capabilities are all interwoven to make the "available," acquirable.

For example, the Miskito are extremely skilled and proficient in locating biotopes in the forest or at sea which will provide the best chances for acquiring meat. Out of sight of land, simply finding a turtle bank presents difficulties. It may be relatively easy if the sea is calm and the turtles are feeding. Cut turtle grass floats to the surface. On the other hand, if the sea is choppy or if the turtles are not feeding, detecting an individual bank brings out the best location-finding abilities of turtlemen. Time since departure, direction of the swells, wind direction, and the "feel" of their *dori* as it rides up and down on the swells over the 10–20-fathom depth banks help the Miskito locate the banks. Tasbapauni turtlemen each know the location of some 20 major turtle banks and twice that number of sleeping shoals; all of which, of course, are underwater and many out of sight from land. A turtleman told me that the sea location-finding ability of the Tasbapauni men was a "gift into the people," but it is more than that. It is the result of an intimate relationship between the turtlemen and green turtles, a cultural relationship built up over hundreds of years of adaptation, and a personal relationship composed of years of an individual's experience at turtling.

Similarly, when hunting in the forest, the Miskito have a directional ability based on their mental maps, which is amazing to an outsider. Bell (1899) commented on this from his experience hunting and tramping the bush with the Miskito:

> In travelling through the bush the Indians are always glancing up at the sun, or observing which way the clouds are drifting, which in this country is always from north-east to south-west. I believe they intuitively keep in their minds a chart of the route they have travelled, and so have a mental picture of the direction they came from, besides which they know more or less intimately the physical geography of the whole country in their district. They know where every creek enters the main river, and they have a fair knowledge of the course and direction of the creeks, the character of the country, and the variety of the vegetation on the banks [pp. 134–135].

The availability and accessibility of food resources operate as adaptive and regulatory forces. Much of Miskito subsistence activities and exploitative patterns are adapted to the resources' temporal–spatial occurrence. The Miskito also are able to regulate the tempo and tenor of their subsistence system by judicious selection of certain food resources over others, thus assuring fairly dependable high yields from land and sea, requiring no large cooperative labor force. Free from time-consuming responsibilities in the village and sharing agricultural labor obligations with women and small reciprocal labor exchange groups, men have the mobility to range widely and often in search of meat for subsistence, or forest and sea resources for market.

CHAPTER VII

PROCUREMENT AND
PRODUCTIVITY

Indian him don't stop for notin'. All dem want is to get dat belly full.
Dem must get dat food, rain or no. Must work to eat.

—A Miskito in Tasbapauni

Much of our knowledge of Indian subsistence systems in Latin America is based on scattered ethnographies and geographical field studies offering good descriptions but providing little in the way of accurate measurement of subsistence productivity, yields, labor inputs, caloric inputs and outputs, and time and distance factors (Nietschmann 1971b).

In this chapter the means of food procurement and productivity in Tasbapauni are analyzed. Emphasis is placed on labor inputs, foodstuff yields, and on the distance to and relative location of specific food-producing areas. Stress is focused on hunting and fishing systems and their productivity. It has been shown how cultural and ecological factors influence the availability and accessibility of food resources. Now it remains to see how those food resources are used, how much effort is involved, and how much food is obtained.

The seasonality of plant and animal resource occurrence is a result of the interplay between natural communities and environmental conditions. In order to resolve the spatial and temporal differences in availabil-

129

ity, procurement and production, activities are scheduled so as to ensure returns from diverse species and sites. The subsistence system of the Tasbapauni Miskito represents a careful orchestration of exploitation patterns operating in phase with overlapping seasonal cycles for fish and game and agricultural resources.

Labor inputs have to be adjusted and scheduled to yield dependable returns within ecologically defined periods and places. The maintenance of the Tasbapauni population, as with any other group, requires a fairly regular throughput of energy, including protein. Because animal protein sources are seasonally diverse and varied, exploitation patterns are regulated between land and water, hunting and fishing, and site and season. Fluctuations in resource availability encourage diverse, wide-spectrum procurement and production efforts within a large area, thereby avoiding year-round pressure on particular species in particular places.

AGRICULTURE

The Agricultural System

You is sure on de shore.
—A Tasbapauni Miskito

The Miskito of Tasbapauni practice a forest–fallow system of shifting cultivation in which the cropping period for any one field is shorter than the fallow period. Their agricultural system is a seasonally and culturally regulated sequence of procedures and events involving the clearing of fields by cutting and burning, planting for 1 or 2 years, and then abandoning the fields until secondary forest cover and soil fertility have been restored. The growing season is yearlong, and water and sunlight are abundant and conducive to maximum growth potential; therefore essential factors for stable maintenance of the agricultural system are land availability, labor arrangements, timing of the agricultural cycle, and the composition and structure of swiddens.

Traditional Miskito agriculture is a complex and ecologically conservative system which closely simulates in morphology and function the tropical forest which it replaces (Fig. 29). The swiddens are man-created models of ecological diversity in species, and in three-dimensional zonation of polycultural plants which maximize utilization of available sunlight, moisture, and humidity while protecting the easily degraded soil from exposure to sun and precipitation. The intermixture of several cultigens with different nutrient and environmental requirements in the same field also serves to protect against plant disease, which spreads

Fig. 29. Structural similarity of swidden and secondary rain forest, near Tilba Rapids, upper Río Coco, August 1969.

more easily in monocultural fields. In addition, the practice of staggering the harvest of swidden crops over a long period of time lessens the detrimental effects on the soil from heat buildup from sunlight, or erosion from heavy precipitation.[1]

1. Detailed explanations of swidden agriculture which emphasize its ecological fitness are presented by Harris (1971), Geertz (1963), Meggers (1971), and Rappaport (1971).

Several factors and processes are at work serving to encourage alteration of the traditional swidden in Tasbapauni. Changes are taking place
in cropping patterns and mixtures, size of fields, fallow periods, labor
arrangements, and in the self-sufficiency of the system. These will be
examined after a generalized discussion of Tasbapauni's typical swidden
agriculture.

Tasbapauni's land is held collectively by the community and available
to any member. Small portions of land have been sold to outsiders and
to a few villagers for coconut groves. These holdings are located north
and south of the village along the beach beyond the villagers' agricultural
subsistence areas. Property rights to a particular piece of land belong
to the individual or family currently working it, or to the person or
family who last worked the field if it is in fallow. To ensure that tenure
rights to a piece of beach land will be respected through the fallow
period, long-living coconut trees are usually planted. Rights to use particular areas remain in the family. The user cannot sell the land to
anyone, but he can renounce his claim, thereby allowing someone else
to work it.

With regard to the limit of potential agricultural land which can be
worked under the existing sociocultural and technological patterns, it
can be said in Leed's (1961) words that

> the proportion of land held in reserve for future cultivation must be several times
> greater than that under cultivation at any given moment, the proportion varying
> with such factors as the rate of fertility recovery, the type of weed and grass invasion,
> the rate of growth of the secondary forest, the rate of mechanical reconstruction
> of the soil, and so on. Thus of the total amount of land potentially available in
> a given expanse of territory, only a part can be horticulturally exploited at any one
> time. Where the total potentially arable land in a given expanse is itself only a small
> proportion of the total area, the amount of land available for cultivation at any
> one moment is, of course, minute [p. 18].

The Miskito consider arable land potentially available for agriculture
if it is located within walking distance of the village (usually less than
6 miles), near a creek in the tropical forest, or alongside a lagoon or
river. Other than the beach plantations, all other field locations are
determined in large measure by their degree of accessibility by *dori*.
Moreover, internal population growth in Tasbapauni and changing
economic emphasis is creating an increasing demand for nearby agricultural land, forcing more intensive use of beach lands rather than traveling
extremely long distances to less used agricultural sites.

Labor arrangements for agricultural activities are of three main types:
division of labor by sex, exchange labor, and wage labor. In making
a plantation, men clear the site, burn the dried plant debris, hoe and

dibble the fields, and help some with harvesting and transporting the crops. Women clean up the plantations after burning, help with planting, harvesting, and transport of harvested crops. Women used to participate more in agricultural work than they do today; the change coming as a result of the various churches emphasizing that a woman should be home with the children rather than in the swidden and because women are turning more to cash economic activities in the village to supplement or supply income for the family. Some women, even married women, have their own plantations which assures family subsistence.

Reciprocal exchange labor (*pana pana*) among friends and families for clearing and planting is limited to subsistence crops and fields. Cash crops (rice and coconuts) are cultivated mainly by wage labor in addition to the operator's labor. The Miskito believe that if a man is going to make money from agriculture he should have to pay for labor. If on the other hand, a man cultivates primarily for subsistence, he will receive help from fellow villagers, for a man must eat and all must help.

The agricultural cycle begins in January or February with the selection and clearing of a new plantation (Table 15). A site is chosen, preferably

TABLE 15

Annual Cycle of Agricultural Work on Subsistence Crops and Cash Crop Rice, Tasbapauni

Subsistence Crops	Month	Rice—cash crop
Begin cutting underbrush and cutting dense forest	Jan.	
Begin cutting secondary forest	Feb.	first burn of rice land if swamp area
First burning, dense forest		
Burning, planting	March	cut down new rice land, second burn, or first burn
Burning, planting	April	cut down old rice land, burn, plant rice
Finish planting	May	plant rice
Weed cassava, rest of plantation	June	first weeding
Weeding	July	rice weeding
Weeding	Aug.	weeding, harvest of 4-month rice
Begin harvesting dasheen	Sept.	harvest 5-month rice, dry and hull rice
Begin harvesting *duswa,* cassava	Oct.	harvest 5- and 5.5-month rice, dry and hull rice
Harvesting	Nov.	harvest 5.5- and 6-month rice, dry and hull rice
Harvesting	Dec.	finish cutting rice, seed rice, dry and hull rice

in secondary growth, as mature tropical forest is too difficult and time-consuming to clear. During my fieldwork period not one new plantation was made in mature forest. Present-day cultivators are benefiting from past clearing of mature forest done by their fathers and grandfathers for banana cultivation. Few banana companies operated their own land-holdings; instead they bought from local cultivators who expanded banana acreage to meet the rising demand.

Since most fields are reused after a fallow period under secondary forest and rights to use a particular field stay in the family, site selection usually involves little evaluation of land capabilities. If the land "gave" before, it will again. Plant indicators of agricultural suitability are used to help decide on where to cut a plantation in old secondary forest. Land is considered to be either "cool" or "hot." Cool land is good and is indicated by the occurrence of such plants as silkgrass (*Aechmea magdalenae*), *waha* bush, and wild plantain. Hot land, thought to be bad for cultivation, is identified by *kira* (*Guazuma ulmifolia*), *atak* (*Geonoma?*), and various cutting grasses. Soil color or composition is generally not overtly taken into account when selecting a site.[2]

The size of plantations varies from family to family depending on family size and responsibilities. Based on plot sizes for a sample of 10 families in Tasbapauni, an average family of six or seven will have a beach plantation of 1½ acres for subsistence crops, one-half acre of coconuts, and a river plantation of 1½ acres for market rice (1 acre equals 2 *tas*).[3] Old plantations in various stages of regrowth will also be used. One family, for example, might have rights to 10 or more contiguous *tas* of beach land to the south and to the north of Tasbapauni, as well as 10 *tas* or so of land on the other side of the lagoon near a river or creek (Fig. 30). Family plantation sizes range from one-half acre to 6 acres depending on size of family, resources, size of kin group and responsibilities, and degree of participation in other food-getting or market activities.

2. Soil types and the effect of swidden agriculture on soil fertility will be presented in another paper to appear later by this author.

3. Precise field size data for all plantations in Tasbapauni were difficult to obtain because of measurement problems of irregular fields and the number of plantations scattered over a large area. Sample plantations were measured with a tape, and an average size was determined. In Tasbapauni, new plantations are often marked off in plots of 50 yards on a side, called a *tas* (1 *tas* equals 0.518 acre, 2 *tas* equal approximately 1 acre, and 4 *tas* equal 1 *manzana*). Wage labor work allotments are by *tas* for clearing. Dividing a field into a number of *tas* also facilitates planting activities and permits the Miskito a certain amount of foresight as to land–crop requirements for the year. The use of the *tas* measurement is also common on the Río Coco (Helms 1967:242). In some areas of the coast, Old Cape for example, fields are not measured, but calculated by the amount of seed and plant bits used.

The Miskito word *tas* probably is derived from the English word "task," that amount of land which could be worked by one man in one day—his task.

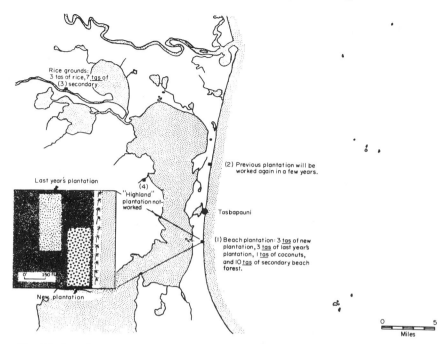

Fig. 30. Location of agricultural areas held and worked by one family, Tasbapauni.

An agricultural work day in a plantation begins at 6:00 or 7:00 A.M. and lasts until 11:00 A.M. or noon, for an average of 5 hours. If necessary, work is carried on to sundown, but work after midday is hot and oppressive; the sun, unimpeded by tree cover, burns down on the cleared fields.[4] One to three additional hours are needed for round-trip travel to a beach plantation, and from 3 to 16 hours or more if the plantation is on a lagoon or up the Kuringwas River.

In clearing a new plantation (*insla disang*), all work is done by axe or machete depending on the amount of underbrush and size of trees. The longer an old agricultural site is allowed to fallow, the greater the overhead shade cover which reduces the amount of underbrush found on the ground. After a 5-year fallow period, saplings have a diameter of 4 or 5 inches. When cut, tree trunks and limbs are chopped into 8- to 10- foot lengths and stacked in piles. Clearing continues into March and April for subsistence plantations and into May for rice plantations. Ideally the cut vegetation should dry 4 weeks or more before burning.

4. For example, at 10:30 A.M. on 20 May 1971, temperature in the shade was already 88.5° F in a plantation, 100 yards from the beach.

After the plant debris has dried sufficiently it is set afire. Practically all smaller limbs, branches, and underbrush are burned, leaving scattered blackened remnants of tree trunks, stumps, and large limbs. The apparent helter-skelter appearance of the newly burned field with its tangles of charred debris, thick ash cover, and partially burned trunks, may obscure important ecological factors. The ankle-deep ash layer helps hide newly planted crops from birds and other animals and increases soil fertility. At the same time, ashes and scattered limbs over the soil help protect it from strong dry season solar isolation and the erosional effects of the first rains falling on the new fields.

Most subsistence crops are planted intermittently throughout the year, with a well-defined planting focus from March through May. Since most of the first yields of these crops would not be harvested until October or November, the traditional Miskito agricultural system called for the cutting and planting of a second plantation in September (*wis nasla*) to supply foodstuffs between May and October. Very few families in Tasbapauni still plant a *wis* ground, due to competing labor demands at this time of year from cash-producing activities such as the beginning of the rice harvest, optimum turtling conditions, and the fact that if they did plant, and had abundant plant foods the next year, they would be socially obligated to share with kin who did not plant a *wis* ground. The conflicts resulting from the increasing priority of cash-oriented activities over subsistence ones will be discussed in the next chapter.

At present then, the most important planting period occurs from March through May. Subsistence crops are planted first, then rice. Miskito polycultural swiddens are characterized by several cultigens,[5] interplanted and growing in the same field along with some residual wild plants which are encouraged because of some potential use. Rice intended for sale in Bluefields is usually planted in one continuous field and not intercropped. Since all cash crop labor outside of the nuclear household is on a wage basis, it is necessary to have the rice lands divided into *tas* for work parcels.

The Tasbapauni Miskito and people in many other coastal villages practice what I call "beach–ridge agriculture." Despite the generally poor sandy soils, the beach areas are important agricultural sites because of the rolling ridge-and-swale topography of series of raised beach bars (Fig. 31). The alternating dry ridges and intervening wet areas, present spatially condensed microecological zones which permit the interplanting of several crops with different water budgets. For example, crops that do better on the drier sandy ridge, such as cassava, *duswa*, bananas,

5. See Appendix B, for a list of plants cultivated in Tasbapauni.

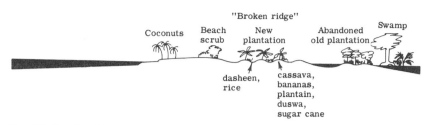

Fig. 31. Beach ridge agriculture, Tasbapauni. A cross section of beach topography showing the alternations of cropping patterns coinciding with beach ridges.

plantains, *plas*,[6] and sugarcane, are placed along the ridges, while dasheen and possibly some rice for family consumption are planted in lower wet zones.[7] Plants that do well in wet soils such as dasheen can be planted a few feet from plants that prefer drier soils for optimum growth, such as cassava. Water-logged soils will rot the cassava tubers; yet in the beach ridges the plants' tap roots can find adequate water from the high water table. Viewed from the air, beach plantations display a striated vegetation pattern, following the microtopography of the surface.

The moon phase is closely watched for planting. The general rule is to plant all aboveground plants from the first quarter to the full moon and all belowground plants from the full to the last quarter. This is said to keep crops from rotting and will ensure high yields. Therefore, at least two planting trips must be made for the same piece of ground.

Regardless of size the traditional polycultural subsistence plantation usually contains the same relative percentages of crops, reflecting not only Miskito taste preferences, but an ecological adaptation to daily food production requirements, food storage problems, and differences in soils. Because cassava is the most esteemed vegetable food and is high yielding, it is planted in the greatest quantity and the most effort is devoted to its cultivation. A subsistence plantation, for example, will contain 1 *tas* of cassava; and 1 mixed *tas* of *duswa*, bananas, plantains, *plas*, and dasheen; with an additional *tas* of coconuts for home and market (Fig. 32; see also Table 18, p. 147).

Coconut palms line the beach to the north and south of Tasbapauni as they do along much of the coastline of eastern Nicaragua. The coconut

6. *Musa sapientum*, a thick, square-shaped fruit looking much like a plantain.

7. Beach ridges have a similar function to ancient ridged fields found elsewhere in Latin America, elevating crops in seasonally inundated areas. See Parsons and Denevan (1967) for additional information on ridge fields.

Fig. 32. Air view of beach subsistence plantations, 3 miles south of Tasbapauni, September 1969. Behind the coconut-lined beach, 1- and 2-acre plantations have been cut from the hone-palm-dominant secondary forest. Various stages of regrowth are evident.

plantings (called "cocals" in Creole) are located just beyond the high tide zone and usually do not extend inland farther than 20 or 30 yards. Coconuts are planted approximately 20 feet apart, 40 trees to the *tas*. A coconut palm takes a minimum of 5 years before it will bear any nuts. A 10-year old, "good bearing" palm will produce 15 nuts every 90 days. Many palms produce for 30 years or more. The nuts are either picked by hand from bunches 20–75 feet overhead or gathered from the ground after they have fallen. A foot loop made from *sani* bark is the only aid the Miskito use to climb the coconut palms. "Drop nuts" are said to yield more oil but take 1 month after reaching maturity to fall. Most Miskito pick coconuts by hand rather than make frequent trips to their cocals to gather drop nuts.

Planting, along with clearing, requires the greatest amount of labor input. Working on a labor exchange basis a one *manzana* subsistence plantation can be planted in one morning with 12 men and 24 women and a number of children (called the "bramble crowd" in Creole English because the children can go into partially burned tangles and plant). The owner is responsible for supplying food to the workers, usually breakfast and the midday meal. Wage labor is paid either by the *tas*, day, or per pound harvested depending on the nature of the job.

Weeding and rice "supplying" (filling in bare spots) take place from June through August. The Miskito say high yields can be obtained on even poor land if the plantation is kept clean of weeds.[8] In weeding, coconuts, cassava, and rice receive primary attention; the latter two are weeded twice before harvest and coconut grounds three or four times during the year. During July, when weather permits, visits are made to the rice grounds on beach lands and river and lagoon plantations to fill in the sparse or bare spots.

Protecting the maturing crops against animal pests is a major chore associated with cultivation. Special visits are made often to the plantations to scare off birds and other animals. White-tailed deer browse on cassava, *duswa*, dasheen, and yam plants (*Dioscorea* sp.); pacas eat the cassava roots; collared peccaries, agoutis, and tapirs can destroy large amounts of crops overnight; parrots and macaws eat ripening fruit. Perhaps the most troublesome pest is the leaf-cutting ant (*wiwi*), which has a penchant for cassava, mango, and breadfruit leaves. Leaf-cutting ant nests are continuously dug out or burned. To curtail crop losses, hunters focus on new and old plantations for white-tailed deer and paca. In fact, a good hunter will usually be able to recoup energy losses from travel and hunting efforts and the destroyed crops by killing occasional deer in the fields.

During the early cultivation period before crops have matured, vegetable foodstuffs are harvested from the previous year's plantation (*insla prata*) and from 2-year old plantations. These crops are of two types: *bila*, or crops replanted for secondary harvest in semiabandoned plantations; and *tukla*, or old crops left in the ground (cassava, *duswa*, yams). Bananas and plantains can be harvested and planted any time of the year, but most are planted in March and harvested in the following year from January through April. Harvesting of ground provisions begins in September with dasheen and in October with cassava and *duswa* and continues on into the next year. Staggered replanting and harvesting tends to spread out the availability of agricultural crops throughout the year, reduces storage and preservation problems, and major crop having a definite harvest time, depending on the type of rice planted (different strains are designated by the number of months to maturity; the most common is "five months").

Crops are harvested intermittently according to need, distance from the village, and time of year. If crops are intended for home use or distribution among kin, visits are made every few days to the plantation to dig out cassava, *duswa*, or dasheen, or to cut a stem of bananas. Root

8. Weeding may benefit crops by reducing competition for nutrients but it may also increase the detrimental effects of sun and rain by exposing the soil.

and tuber crops are left in the plantation until needed—the ground acting as a natural "root cellar" for the Miskito. Because many crops will not keep long out of the ground and because bananas and plantains are preferred green, harvesting is of small amounts over a long period, with much time spent traveling to and from plantations. Women still do a great deal of the harvesting, but men are now helping more than previously (see Conzemius 1932:39). Harvested loads (40–60 pounds) are carried out in burlap sacks from the beach plantations to Tasbapauni. Fewer trips and larger harvests are possible with visits by *dori* to lagoon and riverside plantations.

The number of years of successive cropping in the same field and the length of the fallow period, are interrelated with agricultural yields, the degree of human pressure on the land, and the ability of the soil and forest system to reestablish itself. Some beach plantations are planted 2 or 3 years in a row. However, the general tendency is for a new plantation to be cut each year and the old one semiabandoned. Cassava, dasheen, *duswa*, bananas, and plantains continue to be harvested and replanted, but with very little effort given to weed control and guarding against animal pests. The old plantation is quickly colonized by weeds and various species of secondary growth, especially balsa, palms, and later, *Cecropia*. After a 4- to 5-year fallow period for the much used beach lands, and 5–10 years for other areas, plantations are cleared again. The shorter fallow period for beach plantations is a result of the pressure for frequent use of these fields because of their nearness and accessibility to the village, not because of superior soils.

Reductions in crop yields due to invasion of pioneer species and decline in soil fertility are tolerated up to a point rather than relocate the field and invest additional labor in clearing and planting. Thus, no field is completely abandoned after the first year, but gradually transforms from a swidden site to secondary forest, all the while yielding some foodstuffs. With successive plantings and weedings it is evident that some crops will show marked decreases in yields while others appear to be little affected. For example, 2 *tas* of beach land planted in rice 3 years consecutively yield approximately 1800 pounds the first year, 1200 the second, and 800 the third. Dasheen yields per plant decline similarly: 8–10 pounds per plant the first year, 5–6 pounds the second, and 1–2 pounds the third year. Yet cassava can be grown in the same beach plantation and give rather stable yields for 3 years.

Inputs and Outputs in Agriculture

Obtaining long-term information on labor requirement and agricultural yields was one of the major difficulties encountered in the field research, as most crops, such as cassava, are harvested over many months and old plantations continue to yield for 1 or 2 years after abandonment. Also, acquisition of exact data on labor inputs would necessitate daily field observations on a number of sample plantations during the agricultural cycle; this was precluded by other research activities. Instead, visits were made to plantations during high labor load periods (for example, cutting and planting) in order to measure and record time expended in the task. Sample plots 20 by 20 feet in a plantation were dug up and crop weights recorded. From these samples and from informants' statements[9] an approximation of agricultural labor and yields was obtained. These figures represent "averages" for the most part. Crop yields are highly variable depending on local soil conditions, loss due to animal pests, distance from the village, and intensity of care. Most of the following discussion refers to data derived from beach plantations, the primary focus of subsistence agriculture.

The examples used pertain to 2 *tas* fields (approximately 45,000 square feet) which are very close to an acre (43,560 square feet) in area. The typical family of seven has 3–4 *tas* of subsistence crops, 1 *tas* of coconuts, 2–3 *tas* of rice, several *tas* of land in fallow, one or two of which are still visited for residual crops.

Labor–Time Inputs. Data on agricultural labor inputs were noted as units of time. That is, 1 hour of energy expended in weeding by hand or by machete equals 1 hour regardless of the type of labor. Obviously this is very crude; all agricultural tasks involve different amounts of energy expenditures. Ideally, labor inputs should be measured in terms of calories consumed in agricultural operations such as was done by Hipsley and Kirk (1965) in New Guinea. Lacking energy expenditure tables for the Miskito, I was forced to designate labor as being equal to time.[10]

The number of man-hours spent on work requirements in preparing, cultivating, and harvesting approximately 1 acre (2 *tas*) of beach land

9. Local estimates of weights, time, and distance had to be taken with some reservation. The coastal Miskito use English in counting, distance, and weights because of the unwieldiness of numeration in Miskito. Distance along the coast and at sea is usually expressed in miles, while for rivers and in the tropical forest, hours are used. In Tasbapauni, people's estimates of weights were found to be quite accurate when checked. This ability is probably learned through selling by the pound such things as rice, calipee, and hawksbill shell to merchants. Time estimates, however, were much less reliable. The 5 hours supposedly required to cut down underbrush on 1 *tas*, often turned out to be 10 when recorded.

10. According to Brookfield (1968) time itself is a "valid common measure by which we can quantify labor inputs, traveling distance. . . . In viewing the interconnections of the ecological system and the social system, in assessing the impact of changes in social, economic, and natural factors, we are often only making inspired guesses without data on the allocation of time. Yet solid data on this aspect are scanty and inadequate, even from work studies carried out in Europe [p. 434]."

in subsistence crops and 1 acre of riverside rice land is shown in Table 16. The 602 man-hours spent on the beach subsistence plantation included 26% of that time in walking back and forth from village to plantation. Similarly, 50% or 250 man-hours out of 494 total hours were spent in *dori* travel from the village to the sample rice plantation 1·7 miles distant. The number of visits to scare off animals and to weed and fill in bare spots were reduced because of prohibitive distances.

The total man-hours involved in exchange and wage agricultural labor for a limited sample of three characteristic families is shown in Table 17. Of the 602 hours spent on 2 *tas* of subsistence crops, 132 hours were by exchange labor. Theoretically, these have to be repaid by working on the other plantations. Over one-half of the time spent on rice was from wage labor workers. For one family, approximately 984 man-hours for all types of agriculture were from the nuclear household, 132 from exchange labor, and 291 from wage labor. All estimates include travel

TABLE 16

Estimated Annual Man-Hour Inputs in Miskito Ariculture, Tasbapauni October 1968 through September 1969

	Subsistence plantation		Rice plantation	Coconut ground
	New plantation	1 Year old plantation		
Location	Beach 3 miles S	Beach 3 miles S	Riverside 17 miles NW	Beach 3 miles S
	2 *tas*	2 *tas*	2 *tas*	1 *tas*
Area	45,000 square feet	45,000 square feet	45,000 square feet	22,500 square feet
Activity				
Site selection	3	—	3	—
Clearing	58	—	50	—
Burning	2	—	4	—
Picking up	30	—	—	—
Planting	75	—	12	—
Replanting	15	25	10	—
Guarding	60	10	15	—
Weeding	65	10	105	25
Harvesting	120	75	45	71
Travel on foot	158	50	—	25
Travel by *dori*	16	8	250	12
Total	602	178	494	133

TABLE 17

Agricultural Labor Inputs, Subsistence and Cash Crops[a]

Activity	New plantation NH[b]	New plantation EL[c]	Old plantation NH	Rice plantation NH	Rice plantation WL[d]	Coconut ground NH	Coconut ground WL	
Site selection	1.5	1.5	—	1.5	1.5	—	—	
Clearing	14.5	43.5	—	13.0	37.0	—	—	
Burning	2.0	—	—	4.0	—	—	—	
Picking up	30.0	—	—	—	—	—	—	
Planting	20.0	55.0	—	12.0	—	—	—	
Replanting	15.0	—	25.0	10.0	—	—	—	
Guarding	60.0	—	10.0	15.0	—	—	—	
Weeding	65.0	—	10.0	35.0	70.0	25.0	—	
Harvesting	120.0	—	75.0	15.0	30.0	62.0	9.0	
Travel by foot	126.0	32.0	50.0	—	—	25.0	—	
Travel by *dori*	16.0	—	8.0	114.0	136.0	4.0	8.0	
Total	470.0	132.0	178.0	219.5	274.5	116.0	17.0	*Total*
NH	470.0		178.0	219.5		116.0		*983.5*
EL		132.0						*132.0*
WL					274.5		17.0	*291.5*

[a]Approximate number of annual man-hours provided by nuclear household, exchange labor, and wage labor for same sample plantations as shown in Table 16, Tasbapauni.
[b]NH is nuclear household.
[c]EL is exchange labor.
[d]WL is wage labor.

time. With the exception of coconuts, man-hours spent preparing harvested crops for sale or for consumption are not included.

Since travel time is the major variable in agricultural labor inputs, annual man-hour totals for average size subsistence and cash crop plantations can range from about 1100 to 1800 depending mainly on the distance to the plantations. An increase in size or number of plantations, would, of course, add considerably to the annual man-hours. Excessive inputs of travel time, however, tend to keep down the size of the plantations. Only limited time is available for agricultural work in the fields before one must start back for the village. Large, distant fields require overnight visits or dry season camps; thereby, cutting down frequent village to field travel time.

The total of 602 man-hours for approximately 1 acre is relatively low when compared with estimates available for other subsistence groups. Conklin (1957:150) estimated that the Hanunóo (Philippines) spent from 2865 to 3180 hours per hectare, or roughly, 1160–1287 man-hours per acre excluding travel time. Their yields per unit area, however, are probably high. Lewis (1963:155) gave 113 man-days (1064 hours) for 1 hectare of corn, under hoe cultivation in Tepoztlán, Mexico as being typical. This works out to approximately 430 hours per acre, not including travel time, and for a crop which is harvested all at once. These two examples lend support to the fact that the agricultural system of Tasbapauni is based on very low labor inputs per unit of land and that much of the subsistence effort goes into other activities.

Agricultural Location and Distance. The factor of distance affects the Miskito agricultural system in many ways. The Miskito select the location of their plantations on the basis of a waterside situation, agricultural potential, and distance from the village (Fig. 33). Agricultural areas are not spread evenly throughout any one area or over distance. Instead they are clustered in specific areas. This is for a number of reasons: because the site best fits the cited criteria, and because the Miskito like to plant in areas where other Miskito villagers are planting, thus facilitating exchange labor obligations, and increasing the chances that someone

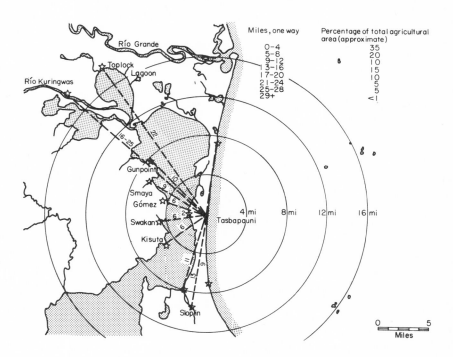

Fig. 33. Distances to agricultural areas. Numbers on lines show actual distance traveled in miles; circles depict straight-line distances.

will be in the area to scare off crop pests or to help in case of an accident (snake bite or machete wound).

The exact location of plantations and the intensity of labor inputs and yields are undoubtedly influenced by increasing distance from the village. Although not enough quantitative data were obtained on agricultural inputs and outputs from samples to make a definitive statement on the exact nature and degree of this influence, a rough correlation can be described. Actual daily work time in fields is diminished with increasing distance from Tasbapauni up to about 10–15 miles by water, beyond which overnight camps must be made. Crop yields are lower for distant plantations because of less weeding and losses due to unmolested crop pests. Harvests are fewer, but amounts per trip are larger than for less distant fields. *Dori* loads of agricultural foodstuffs cause storage problems for the cultivator; much has to be distributed in the village or it will spoil, and in the end the family gets less. Close beach plantations are preferred because they can be visited often and at any time of the year; whereas, unfavorable winds and storms may keep an individual from a plantation across the lagoon for days.

A family may have two or three small plantations in different areas to spread out subsistence risk by minimizing possible crop loss due to flooding, animal pests, or disease. Large subsistence fields, beyond the size and productivity needed for nuclear family food supplies and extended family-kin food distribution, are not common, nor are they to be encouraged in the Miskito's view. Miskito exchange labor arrangements function within a range where everyone's fields are approximately equal relative to family size and kin obligations. Beyond subsistence size, wage labor is required. Money is in short supply. Furthermore, Helms suggests that the Miskito operate on the basis of attempting to restrict situations that might lead to loss. "If an individual loses, someone else gains at his expense. Hence one should protect against loss [Helms 1967:412]." A Miskito does not like to think that someone else outside of his kin network will benefit from his extra work. It is better to plant small plantations than to risk losing crops to animals or to "thieves." Or as one Miskito put it: "I no going to work to feed dem other people." A Tasbapauni Miskito would rather use time above minimum agricultural labor inputs to go turtle fishing or hunting than to become encumbered with a lot of land and plantings.

Agricultural Yields. One of the most important aspects of an agricultural system is, of course, its productivity. This can be determined by measuring the human energy inputs and foodstuff yields per unit area, or the yields per hour of labor input or per unit of land.

In discussing the agricultural productivity of Tasbapauni plantations it is important to note that "the best general rule to the behavior of primitive farmers is that they work to get the maximum return for the minimum effort [Nye and Greenland 1960:129]." The Miskito practice "hurry-up" agriculture exclusively; all agricultural tasks are accomplished rapidly, with little delay and few rest periods. The Miskito say all things will grow with little human effort; yields are thought to be determined for the most part by how well the plantation is kept free of weeds.

The total yields of traditional crops from a sample plantation are presented in Table 18. Yields for the second year were obtained from an adjacent *prata* ground (second year plantation) worked by the same informant. The total yields in pounds were converted into calories to achieve some degree of commensurability. Calorie yields were also estimated for a rice plantation (hulled rice), coconut ground, and mango and breadfruit crops from trees next to the informant's house in the village.

An estimated 2,275,000 calories were harvested off the 2 *tas* new plantation and 1,886,000 from a second year plantation. In addition, from 1 *tas* of coconuts, producing some 2400 nuts annually, 10% were used in the home for cooking oil, providing 198,240 calories. Almost the entire rice crop was sold, with only about 5% or 122,395 calories being consumed in the home. Finally, 109,774 calories came from mangoes and breadfruit. The total produced on the seven-member family's subsistence plantation and from subsistence portions of rice and coconut plantations amounted to some 4,600,000 calories. Market-bound rice and coconuts totaled 3,777,385 calories. Some of the money obtained from the sale of these crops and from coconut oil was used to purchase store foods.

Granting that calories themselves are not the only thing to be looked at in terms of agricultural yields or nutrition (see Chapter IX), they do provide a common denominator for comparing different crops.[11] If we take 2500 calories per day as being the average daily consumption rate for this informant, approximately 900,000 calories are needed per year. With his wife and five children they consume approximately 4,500,000 calories from all sources per year, or about the same as he produces from subsistence agriculture alone. Actually agriculture contributes less than this because a large portion of his subsistence production goes to close kin and friends. Although a large amount of calories is still obtained from hunting, fishing, as well as from purchased foods,

11. See Rappaport (1968:49–50; 280–281) for detailed tables of crop calorie yields for the Tsembaga Maring of New Guinea.

Yields for 2 tas of New and Old (prata) Plantations, 2 tas of Rice, and Miscellaneous Tree Crops, Tasbapauni, October 1968 through September 1969

Crop	Percentage of area	Square feet est.	Plantation New (yield in pounds) 0–12 months	Old 12–24 months	Percentage edible	Number of calories per pound	First year calories	Second year calories	Total calories
Subsistence crops									
Cassava	50	22,500	2000	1650	80	598	974,800	789,360	1,764,160
Dasua	18	8100	400	225	80	598	191,360	107,640	299,000
Bananas, plantain, and *plas*	15	6750	2500	2350	70	498	871,500	819,210	1,690,710
Dasheen	7	3150	425	385	85	417	150,537	136,359	286,896
Sugarcane	1	450	375	300	30	371	41,923	33,390	75,313
Pineapple	2	900	225	—	75	236	39,884	—	39,884
Papaya	1	450	45	(birds ate)	75	145	4930	—	4930
Stumps	6	2700	—	—	—	—	—	—	—
Totals	100	45,000	5970	4910			2,274,934	1,885,959	4,160,893
Rice cash crop									
Rice (6 months)	90	40,500	1500 (hulled)		100	1635	2,451,500		2,451,500
Stumps	10	4500							
Totals	100	45,000							
Tree crops									
Coconuts (10 years old)	one mature tree (in village)	22,500	2400 nuts		100	1341	1,609,200		1,609,200
Breadfruit			400		68	367	99,824		99,824
Mango	one mature tree (in village)		100		50	199 (green)	9950		9950

agriculture is the major source of calories, and its importance is increasing as we shall see, in the face of growing population and declining meat yields.

The importance of the *prata* grounds in the subsistence system of Tasbapauni needs to be stressed. The second-year agricultural grounds supply 40–50% of the calories yielded by a new plantation with only 30% of the labor inputs. In addition, the old plantations are major foci of hunting activity, especially for white-tailed deer which are attracted by the presence of the plant foods. The *prata* forms an important segment of the compact, ecologically diverse, beach provision grounds.

Thus, in a small area there is a range of habitat conditions for wet and dry crops and for desirable game animals. In all plantation areas, whether on the beach, river edge, or "highland," the juxtaposed new and old plantations offer a range of foodstuff yields and labor requirements, as well as sites for obtaining protein from hunting to supplement the largely carbohydrate agricultural foods.

Factors of Change

Land, labor, and crops are increasingly becoming commodities with a value apart from a social or subsistence context. Within the Tasbapauni Miskito's agricultural system the economic is being differentiated from the social at the expense of the ecological. The major factors initiating change within the agricultural sphere are: (1) increased population pressure, (2) reallocation of labor inputs to cash-yielding activities, (3) cash exchange supplanting reciprocal exchange of labor and goods, and (4) the competing values of younger Miskito. The net effect is the dislocation of reciprocal agricultural production and distribution, and replacement by individualization of production by nuclear families. "De way dis place going, everyone moving for demselves," said an elder Miskito. The major factors of change with the agricultural system appear to be occurring in labor arrangements, land use, field size, crop composition, and scheduling.

Labor arrangements for agricultural work are being altered by men replacing women, and with the increasing importance of wage labor in the village. Women contribute less to agriculture than they once did, due to the influence of the churches, as previously stated, and because women are participating instead in the production of coconut and copra in the village for outside market sale. The clearing, planting, and cultivation of cash crops is done with the aid of hired laborers. Where once large subsistence plantations were made and the communal workers rewarded with prepared food and harvesting of all they could carry, today the preparation of a plantation larger than for family subsistence requires cash payment for workers. With available cash in short supply, smaller plantations are being made.

Agricultural land use is being modified as well. More pressure is being placed on the preferred beach lands by the village's growing population. The response has been a reduction in the length of the fallow period and a change in the combination of crops. According to older informants beach plantation fallow periods used to be 10–15 years long while at present they are often as short as three or four, the average being less than ten.

Although it was not measured in the field, the Miskito claim that crops respond differently to shortened fallow periods. Some crops such as cassava and *duswa* can tolerate short fallow without drastic reduction in yields, while other crops such as bananas and plantains, do not do well in frequently worked fields. Outside markets for coconuts (and coconut oil and copra) and rice are expanding which encourage homogeneous plantings. Present-day trends suggest that polyculture agriculture will increasingly give way to monoagriculture in the future and in so doing, lose many of the ecological advantages now found in the polycultural swidden system.

In the traditional Miskito agricultural system, clearing, burning, and planting activities were carefully coordinated to coincide with optimal periods of dryness or rainfall. A major planting time in April and May, recurrent replanting of vegetatively reproduced crops, and a secondary planting time in September spread out crop maturation so that subsistence plant foods were available year-round. However, it is now evident that various competing means of livelihood are siphoning off or creating a rescheduling of subsistence agricultural labor, sometimes to the extent that it is out of phase with environmental conditions. Shrimp fishing, turtling, and rice harvesting provide cash and usually preclude or relegate subsistence agriculture to a secondary priority for time allocation. More to the point, an informant told me, "See, what de problem is, is dat everyone want to live off money."

ANIMAL HUSBANDRY

The raising of animals for food plays a very small role in Miskito subsistence activities. Consumption of domesticated animals is usually limited to times of meat shortages and to December when a cow or two may be butchered in the village for Christmas. Chickens, pigs, and an occasional cow are sold to merchants from Bluefields who make frequent visits to Tasbapauni and other Pearl Lagoon villages for animals.

Pig and chicken keeping are the two most important activities involving domesticated animals. Most families have one or two pigs and a half dozen or so chickens. Pigs and chickens are in effect, self-foraging bank accounts which can be cashed in when money is needed. These animals, occupying the end rung on the food distribution system, subsist largely on coconut "trash" (coconut meat which has been squeezed for "milk" and "cream") and other kitchen wastes. Pigs are fed clusters of hone palm seeds and occasionally green bananas. Human and animal feces are a major source of food for both animals, constituting a very "economical" recycling of food. Because of cultural preferences, the Miskito food system with respect to pigs and chickens is an open one, and most of these animals leave the system to reenter in the form of money which is transformed into such things as clothes, flour, sugar, and coffee.

Young wild animals are sometimes obtained during hunting trips. Parrots, macaws, toucans, and monkeys are kept as pets, while more infrequently young white-lipped peccary are raised until they are old enough to be eaten, or until they are killed by dogs in the village. If taken young enough and held close for a moment after capture, such an animal will become very attached to an individual, following its new "protector" everywhere.

GATHERING

The collection of wild edible materials is generally a secondary food-getting activity. Wild fruits, bird and reptile eggs, and shellfish are sought after by hunters and turtlemen while they are away from the village, and they are gathered as ancillary and emergency foods for the village. Leaves and bark are gathered for teas and medicinal purposes. Some gathered foods form a regular and dependable part of the diet and receive specific attention during periods of availability.

Little success was attained in attempting to measure time and distance inputs and foodstuff yields for gathering. Some gathering and eating of fruits is done in conjunction with hunting, agricultural work, and turtle fishing, necessitating on-the-spot observations of time inputs. In addition, few of the foods gathered by the Miskito of Tasbapauni have been analyzed for their caloric–nutritional contents. Women and children gather much of the wild plant and animal foods at various times of the day, making measurement efforts difficult.

The Miskito give special attention to collecting certain foodstuffs: *ahi* shellfish (*Donax*) from the beach, iguana and hicatee eggs from river sandbanks, coco plums and sea grapes from the beach, *ibo*, nance, and hone fruits from the highland and beach forest, and honey from wherever it can be obtained.[12] The rest of the gathered foods are occasionally brought into the village or are eaten by hunters and turtlemen while away after game (Table 19). Many of the gathered fruits are ferel, located in old agricultural areas. Guava, granadilla, and papaya seeds are commonly spread by birds.

Ahi shellfish (*Donax* sp.) are probably the most important gathered food in Tasbapauni. They can be obtained in the intertidal zone throughout the year, but gathering is easier in the dry season, when the surf is low and the *ahi* are said to be fatter. Women and children collect *ahi*. An 8–10 pound bucketfull takes about 2 hours to gather. After

TABLE 19

Wild Edible Materials Collected by People from Tasbapauni–Forest, Beach, and Cays[a,b]

Hunters			Turtlemen			Villagers-tasbapauni		
Locust	(F)	C	Whelks	(Cay)	C	*Ibo*	(F)	C
Nance	(F,B)	C	Conch	(Cay)	O	Hog plum	(F)	O
Monkey apple	(F)	C	Lobster	(Cay)	R	Sapodilla	(F)	C
Bread nut	(F)	O	Sea grape	(Cay)	C	Nance	(F)	C
Bihu	(F)	O	Coco plum	(Cay)	C	Coco plum	(B)	C
Wild granadilla	(F)	O	Coconuts	(Cay)	C	Hone	(F)	C
Pingwing	(F)	O	Bird eggs	(Cay)	R	Sea grape	(B)	C
Kisu palm	(F)	O	Hawksbill eggs	(Cay)	R	*Ahi* (*Donax*		
						shellfish)	(B)	C
Bribri	(F)	O				Iguana eggs	(B,F)	C
Pigeon plum	(F)	O				Hawksbill eggs	(B)	R
Corozo	(F)	O				Hicatee eggs	(F)	C
Wild cacao	(F)	O				Honey	(F)	O
Iguana eggs	(B,F)	C				Locust bark		
						(tea)	(F)	C
Caiman, crocodile						Chiny root		
eggs	(F)	O				(tea)	(village)	C
Hicatee eggs	(F)	C				Lemon grass		
Honey	(F)	O				(tea)	(village)	C

[a]See Appendix B for identification

[b](F) is forest; (B) is beach; (Cay) is cays; C is common; O is occasional; R is rare.

12. The Miskito go to great lengths to acquire honey. While paddling along creeks, or walking in the forest, they are always on the alert for a beehive. If bees are seen going into a tree, a Miskito will mentally note its location and return in May, when the honey is said to be best. Rather than destroy the comb and bees, they will try to bring the hive intact back to the village, where part of it will be placed in a section of bamboo and hung under an eave to encourage the bees to continue producing honey. This is done with stingless bees (*Melipona*). The Miskito recognize at least 15 kinds of bees. *Waikira* is a stinging bee and *sitsit* is stingless.

2 or 3 hours of boiling, tediously picking the meat out of the shell, and continual rinsing to rid the meat of sand grains, approximately 2 pounds of meat are obtained. *Ahi* gathering, as with with most other collected foods, helps to reduce subsistence risk because they occur in known sites at known times. The Miskito say that when there are no turtle to strike (harpoon) "women have to go strike the *ahi*. Ahi is sure."

The gathered foods are important in that they require no special technology, are often very high in food value complementing the carbohydrate-based diet, and some can be obtained during periods of food scarcity such as July and August. They require no maintenance and only small labor inputs for acquisition. In addition to the wild plants recognized as foods, many wild plants are collected for medicinal purposes. The practice of bush medicine is widespread, and every house has a few dried plants stuck up on a shelf in case of sickness.

In other Miskito villages where biotic composition is different, other wild foods may be gathered, such as oysters, freshwater snails (*suti*), and freshwater shrimp. On the Río Coco gathering is important as a source of emergency foods, but is probably not as diversified in terms of types as in Tasbapauni or other coastal villages where the variety of local environments offers a wide array of wild foods.

HUNTING AND FISHING

Striking turtle, shooting wari, dats what de living is in dis place.
—A Tasbapauni Miskito

Methods and Technology

Hunting and fishing activities form the core of village life. Differences in hunting and fishing methods complement the mosaic composition of the Miskito's land-and-water environment. Just as their environment is made up of different places with various fauna, their hunting and fishing methods are composed of analogous sets of different technologies, strategies, labor arrangements, and sea and forest knowledge. Fishing alone involves a complex set of radically different technologies and strategies for turtle, manatee, shrimp, and different fish. Few can master, as well as afford, the varied technology and necessary skills to be successful in all phases of hunting and fishing, at all times of the year, and with equal proficiency for land and water.

Hunting and fishing are considered basically men's work, women seldom participating except in fishing with hook and line. Hunters go out alone or in groups of up to five or six if a big drove of white-lipped peccary has been reported. Hunting partnerships are rather loose, and different combinations of individuals are common. Turtle fishing, on the other hand, requires very close cooperation between the "captain" or sternman and the "strikerman." Partnerships form around each individual's skill, reliability, and temperament. A turtleman has to have a partner whom he can depend on, who is ready to go whenever the wind changes or the sea goes down.

Meat and fish are distributed equally to all those participating in the activity. The captain and strikerman split 50–50. In addition a meat share is taken out of the catch for the owner of a loaned *dori*, gun, or shrimp net—however the meat was obtained.[13] According to informants, *dori*, gun, and net shares have been only recently introduced into the village. They claim that when game and turtle were plentiful, there was no need to share, because everyone was successful. Now, however, when turtles and game animals are fewer, and village population higher, by sharing everyone has more of a chance to get meat. This adjustment in the village reduces friction over all-important meat.

Out of approximately 160 men between the ages of 21 and 60, 124 did some hunting and fishing during the 1968–1969 period of field research.[14] Of these, 80 turtled (65%), 26 hunted and turtled (20%), and 18 hunted (15%). There is no clear distinction for fishermen as most of the men will occasionally go after fish whether they are hunters or turtlemen. Not all of the 124 hunters and turtlers were active on a year round basis, usually because of the lack of necessary equipment or skill. For example, turtlemen distinguish themselves as being either a "striker" (harpoon and line) or a "net setter." A striker must be knowledgeable as to turtle habits and behavior for he must locate, stalk, and hit them. Thus a striker can operate almost all year in rough sea or calm as long as there are turtle. A net setter, in contrast, is active solely during the long and short dry seasons, and he has but to place his nets in areas of turtle movement and check them every morning. This takes skill, to be sure, but not to the same degree as for a strikerman. It does take money, however, as the materials for one turtle net cost 50 *córdobas* ($7.00) if made from nylon line, and a net setter usually has 10–25 nets.

13. If two turtlemen without a *dori* should borrow one, they must give the owner a front or hindquarter of meat from each turtle taken and three *córdobas* ($.42) from the sale of the calipee. The loan of a gun or shrimp net brings the owner a similar one-quarter share unless other arrangements are made.

14. The remaining 36 men did no hunting and fishing because they were either sick, disabled, had a sustaining specialty craft or occupation (building sea-going dories, repairing boats, diesel boat owners), or simply did not like the rigors of hunting and turtle fishing. They usually were able to obtain meat by purchase or through kin.

Even though the village of Tasbapauni is located favorably for easy access to the tropical rain forest or to the shallow offshore waters, 80% of the men concentrated their activities on either fishing or hunting, with but 20% of the men exploiting both land and water environments for meat. The reasons for the unequal focus of hunting and fishing activities can be attributed mainly to: (1) the dependability and productivity of turtle fishing, (2) the water orientation of the Miskito, (3) prohibitive costs in owning two sets of equipment for both hunting and fishing (Nietschmann 1972).

The 26 men who both hunt and take turtle have mastered the skill and knowledge to operate in land and water environments, but probably more importantly they have the equipment to do so. A good sea *dori* costs from $50.00 to $100.00 to build. A .22 rifle costs $50.00, a single-shot shotgun $40.00, and a repeating shotgun $100.00 (Fig. 34). A cast net for shrimp is $15.00 if labor has to be paid for. Additional outlays for shells, harpoon lines, and *dori* upkeep require considerable expenditures of money. Therefore, necessary equipment costs can be restrictive

Fig. 34. Miskito hunter with his prized shotgun. These guns have greatly increased the effectiveness of hunting white-lipped peccary, which travel in droves, and offer chances for numerous shots.

for individuals who want to do both hunting and fishing. Economic constraints limit involvement in two meat-getting systems.

The importance of the *dori* for Miskito livelihood cannot be over-emphasized. Due in part to the village's haulover location, 98% of the fish and game by weight was brought into the village by *dori* during the first fieldwork period. Eighty-seven percent of the meat was obtained with a *dori*. Thus the *dori* is one of the most important pieces of equipment a Miskito can have. Except for the use of .22 rifles, shotguns, nylon lines, and turtle nets, most of the hunting and fishing techniques and equipment are little changed from those in early 17th century descriptions. It is the *dori* and the harpoon, which still provide approximately 60% of the village's meat.

Tasbapauni dories are known for their fine workmanship, durability (up to 20 years), and seaworthiness. The best dories are made from mahogany (*yulu*), santa maría (*krasa*) and cedro macho (*saba*) logs. Only a few men can cut a good sea *dori* (Figs. 35 and 36). Turtle-striking

Fig. 35. Cutting a *dori* hull from a *saba* log, Tasbapauni.

Fig. 36. Fitting a "stempiece" to a *dori* bow, Tasbapauni.

dories usually have an 18–21 foot length and a 3-foot beam. Hulls are roughed out with hand axes and adzes. Mahogany or cedar (*yalam*) planks are added to the sides, and white mangrove (*laulu pihini*) ribs are fitted to give the *dori* strength and rigidity. The bow is formed with carefully cut and fitted pieces of cedar or mahogany; a curved piece of soft, shock absorbing wood (*moho* or *kahmi*) is used for the "stempost." Dories are steered with paddles, so no rudder is fitted. In addition to labor, costs involve nails, paint, floursacks for sails, and lines for rigging.

The Miskito employ an array of different points and harpoons for striking fish, manatee, and turtle (Figs. 37 and 38). Two types of fish lances with fixed points and a harpoon with a detachable point are used. A special point and harpoon are used for manatee. All harpoon points are made out of metal files. Turtle harpoons are usually purchased from Rama Indians who are able to find the especially hard "wild supa" (*Guilielma* sp.), best suited for harpoons, located in the forest south of Bluefields Lagoon. Orders for a "turtle staff" are sent with Tasbapauni boatmen to be passed on to Rama Indians who visit Bluefields for store goods. This trade relation is a carry over from previous days when Miskito turtlemen, on their way to the hawksbill fishery close to San Juan del Norte, used to stop at Rama Cay (Bluefields Lagoon) for turtle harpoons and points (Conzemius 1929b:305).

Fig. 37. Harpoons and points. From left to right: Barrel ends of a fish harpoon and a turtle harpoon (9 feet long) and harpoons for fish, manatee, and turtle. Scale at right is in inches.

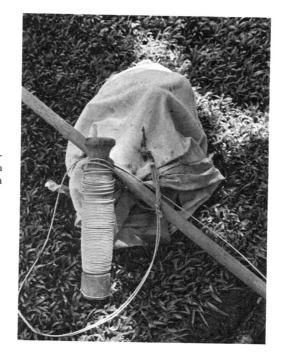

Fig. 38. Manatee fishing equipment. The sack contains a dozen balsa-like bobs, each wound with 10–15 fathoms of line.

Fishing

When de wedder fine, go look a turtle; when bad go walk de bush.

Most of the fishing activity revolves around turtling, while fishing for shrimp, fish, and manatee are secondary and more seasonal pursuits.

The turtlemen usually leave at 2:00 or 3:00 A.M. with the land breeze and by late morning are among the cays 12–25 miles offshore. Out on the banks, both captain and striker are alert for a floating turtle or the hiss sound of a turtle blowing as it surfaces nearby. Once a turtle is spotted, communication between the men consists of hand signals from striker to captain whose responsibility it is to approach the turtle's head or tail—its blind spots. Any distance under 20 yards is considered acceptable for a shot. The harpoon is thrown in a high arc, striking the turtle almost vertically, so that the point will not glance off the carapace. Sometimes the turtles are harpooned as they swim just below the surface, or struck "wind and water" as the Miskito say. The turtle is then hauled in, lifted into the *dori* and put on its back where it will remain motionless (Fig. 39). After a day on the banks, the turtlemen return to the cays where the turtles are placed in coral-walled crawls ("klars") or dragged up on the beach into the shade. Each cay is ringed by a coral reef and fringed with blinding white sand which restores heat and life to feet cramped and wet from the trip and the chase.

The turtlemen will stay out on the cays and banks until they get their load of three or four turtles. If there is a shortage of meat in the village

Fig. 39. Hauling a small female green turtle into a *dori*, September 1969. The turtle harpoon point ("peg") can be seen stuck into the carapace.

or in their house, they will return as soon as they can get a good-sized turtle. Usually the turtle dories return in late afternoon, running before the northeast wind. Villagers line the beach straining to see the white sail specks coming into land. As the dories approach closer, each will be recognized by the cut of its sail and, when closer yet, the number of turtles each *dori* carries will be guessed at from how low in the water it rides. This is a happy time in the village; the day's work is done, the turtlemen are coming in, and there will be meat tomorrow (Fig. 40).

Fig. 40. Unloading turtles on the beach in the late afternoon, Tasbapauni. Turtlemen are dragging one of the two large green turtles harpooned on the banks. A medium-size hawksbill lies in the *dori.* *Supa* palm harpoons and cotton and nylon lines are also visible. The palm leaf is used to shade the turtle.

During the calm weather month of September when the sea is very flat, turtlemen sometimes go out at night "torchin' turtle." The turtlemen wait quietly and tensely in their dories, floating over the shoal and coral head areas where green turtles sleep. When the turtles break the surface for air, the flashlight-armed sternman tries to spot the sound so that the strikerman can make a quick throw with his harpoon.

Fishing is an important subsidiary meat procurement activity, supplementing hunting and turtling returns. Even though fishing is reported to be better in the dry season during low water, this is when most labor activity is focused on turtling and agriculture. The majority of fishing is done during the rainy months when it is difficult to get other meats.[15] The preferred means of taking fish is by striking them with one of many light harpoons or lances. Also, depending on species and habitat, fish are taken with hand lines, seine and cast nets. The Miskito of Tasbapauni, unlike those on the Río Coco, no longer use bow and arrow or poisons (*Seriania inebrians* and others) for fish.

The Miskito are able to distinguish different fish by their wakes. The harpoon is thrown, allowing for depth, deflection and fish species. One observer writing in 1699 noted that the Miskito gathered at the mouth of the Río Coco during the dry season to strike fish with harpoons:

> Which they throw 20 or 30 yards from them at a single fish, which they cannot see, through the thickness of the water, saving only a little curling wave, which they call the fish's wake, and by that they guess how deep he swims under water, it may be two or three foot; in which exercise they very seldom miss their game (M.W. 1732:287).

Shrimping involves a large number of men during the months of April, May, and early June when shrimp enter Pearl Lagoon and congregate on the banks at the northern end of the lagoon. Shrimp buyers come up in ice-laden boats from Bluefields and the town of Pearl Lagoon to buy shrimp. Fifty to 150 dories from lagoon villages may be found on the banks. The shrimp have always been there but only in the last 4 or 5 years has there been a market. The chance for ready money draws men off the turtle banks and plantations. When shrimp are "running," two partners can easily make 50–70 *córdobas* a morning catching and selling 100–150 pounds of shrimp. Shrimping activities among the Miskito are also important in Río Grande and Wounta, besides

---

15. Bennett (1959) also reported the prevalence of wet season fishing over ecologically more favorable dry season fishing among the Chocó of Panama:

> There is a tendency for fish to become more important in the diet of the Indians during the wet season than in the dry season. This is due to the need to stay close to the growing grain crop, the reluctance to remain out in the forest on a hunt during periods of heavy rain, and to an occasional lack of ammunition resulting from a shortage of money before some of the harvested grain is sold [p. 68].

the Pearl Lagoon Miskito villages of Haulover, Raitipura, and Kakabila. Continuous exploitation during the shrimp season in the lagoon over the last few years, coupled with extensive, year-round shrimp boat activities along the coast, has probably drastically reduced the breeding population. In 1971 shrimping was very bad, only a very small amount of shrimp was caught in the lagoon.

Fishing for manatee requires great specialization, patience, and skill. Because manatee have such an acute sense of hearing, special narrow paddles must be used, and *dori* bows must be shaped to create no noise in the water. Manatee are also taken by harpoon but the harpoon and points are different from those used in turtling or fishing and the lines have floats. Of all hunting and fishing practices, manatee fishing involves the highest inputs of time. Many trips to lagoon banks and manatee feeding areas[16] are made before a fisherman is successful. The manatee men say this is because the animals are frightened away by all the noise from concurrent shrimp fishing. Also manatee are said to be scarcer than 20 or 30 years ago. Finally, it is difficult for the skilled manatee fishermen to find and train crewmen to accompany them. Probably the sole reason any manatee fishing occurs today is because of the promise of obtaining so much meat all at once (a full-grown manatee will yield 200–300 pounds of delicious meat).

Hunting. Hunting functions as the main meat-getting pursuit for individuals other than turtlemen and as an ancillary activity for turtlemen when turtle are unobtainable. Favorite hunting spots are old and new plantations when the young plants are just beginning to sprout and in palm swamps and the banks of rivers and creeks. Though optimum hunting conditions generally occur in the dry season, the work emphasis on turtle and on agriculture precludes all but a little hunting at that time. Hunters pay particular heed to seasonal ripening of fruits which attract game animals. The Miskito prefer to hunt in the swamp where they will have a chance at both white-lipped peccary and white-tailed deer—there being few white-tailed deer in the mature rain forest. In the swamps the going is easier, fewer thorns impede the barefoot hunters[17] (Fig. 41).

16. Manatee fishing is sharply focused on specific areas where the animals feed. One experienced informant told me that "de mananti dem, dey is just like a cow, have portrero [field] comn' feed."

17. In the tropical rain forest, thorns present a constant hazard to hunters. Despite thickly calloused feet, some thorns (*kiaya* in Miskito or "prickle" in Creole) can puncture the skin, particularly those of the *kaka* palm tree. Chagnon (1968) while describing the Yanomamö of Venezuela and Brazil, called attention to the hunters' thorn problems in the tropical forest: "The most common and recurrent injury the Yanomamö sustain is thorns in the feet. A party of ten men can rarely travel an hour without one of them stopping to extract a thorn from his foot. . . . Their feet have thick calluses on them, but because the trails frequently cross streams and swamps, the calluses often soften and are easily pierced by thorns [p. 20]." Some Miskito have taken to wearing boots to avoid thorns, but boots are an added expense and wear out quickly.

Fig. 41. Hunting in a palm swamp on the edge of Pearl Lagoon, south of Tasbapauni.

Some hunting is done at night using flashlights to "torch" for nocturnal feeding animals. Even if the Miskito cannot see the animal's outline they can identify it by its distinctive eyeshine (color, spacing, and height of the eyes) reflecting in the flashlight beam.

During the heavy July and August rains hunting supplies the village with meat. In the low coastal and river floodplain areas animals become stranded on natural levees and hillocks trying to escape rising flood waters. Bell (1899) described this hunting technique:

> They surround the islands in the rivers while the flood is rising, and when the water has driven all the animals to the top they land and kill them, pursuing in canoes those that try to escape by swimming. This manoeuvre produces a plentiful supply of agoutis, pacas, armadillos, and deer, but an island yields only one harvest in the year [p. 246].

Hunting and Fishing Focus and Yields

Despite the fact that many indigenous peoples in tropical Latin America still rely heavily on hunting and fishing for their major source of meat, we know very little about the actual significance or effectiveness of those activities, or their impact on the animal populations (Nietschmann 1972).

We have only vague information on the amounts and types of fish and game animals exploited by a particular human population over a period of time. Bennett (1971) emphasized the paucity of available data on hunting and fishing when he wrote:

> The exploitation of the wild animal resources base by Amerinds as a source of food has received inadequate attention from scholars who have studied Amerind societies. Too often a list of animals said to be eaten by the group investigated is offered as sufficient information. . . . One is generally told little or nothing about the quantities of a given taxon taken, seasonal aspects (quantitative and qualitative) of hunting and fishing activities, the nutritional contribution made by such animal foods to human diets and the details of meat and fish storage and/or preparation [pp. 35–36].

Tasbapauni hunting and fishing focus and productivity were major themes of our field research. Although hunting and fishing yields were generally easier to obtain than agricultural yields, data on time and distances involved were not. Some hunting and fishing took place in conjunction with agricultural activities. That is, after clearing, planting, or weeding the men might stay in their plantations to hunt at night or they might hunt and fish from their dories on the way back to the village. The distances recorded in hunting and turtling denote only round-trip miles from the village to where the animal was taken. They do not include how many miles were traveled searching for game.

The recording of numbers and weights of the larger game animals is believed to have been reasonably accurate as their arrival in the village and butchering were quite conspicuous. However, recording small fish and game animals proved more of a problem. Hunters, fishermen, and turtlemen were visited daily, when possible, in order to note any animals taken. Usually informants were able to report on many small catches of fish or game brought into the village. While walking about the village, I also kept close watch on kitchen refuse piles for remnants of an animal: a hicatee carapace, a small jawbone, some feathers. People probably wondered at my rather circuitous routes through the village.

Hunting and Fishing Yields. From October 1968 to the end of September 1969 animals taken in hunting and fishing by Tasbapauni villagers were identified and weighed.[18] Whenever possible animals were weighed

18. A number of scales were used in weighing. For smaller animals 10-, 50-, and 100-pound capacity spring scales proved adequate. These were checked against weights to assure accurate readings. Large animals such as tapir and manatee were handled by cutting them up and weighing the pieces. Green turtles were recorded with the aid of 100- to 400-pound capacity steelyards used to weigh hulled rice. These were "calibrated" with known weight cans of gasoline. After a number of weighings of green turtles, white-lipped peccaries, and white-tailed deer, I found that I could estimate the weight of one by eye within a few pounds. This not only saved much fieldwork time, but saved also on washing blood-splattered T-shirts.

before and after butchering. Butchered weights listed in Table 20 for various common game animals represent "clean meat" portions after the animal had been gutted, cut up, and most of the nonedible parts removed. Bones were included in butchered weights as they are not removed for the sale of meat or for cooking.

There were, of course, considerable size variations for particular individuals taken, depending on age and food supply. Green turtles, for example, ranged from 30-pound live-weight "chicken turtles" to 350- and 400-pound monsters. A medium-size green turtle weighing 190 pounds, say, would yield:

Piece	Weight	Use
calipash	2 pounds 10 ounces	eaten
organs	6 pounds 8 ounces	eaten
meat	84 pounds 0 ounces	eaten ⎫ 100 pounds 10 ounces
front and back fins	10 pounds 4 ounces	part eaten ⎬ butchered meat
head	4 pounds 8 ounces	part eaten ⎭
shells	32 pounds 10 ounces	discarded
calipee	5 pounds 6 ounces	sold
dung	18 pounds 0 ounces	discarded
blood, waste	27 pounds 0 ounces	fed to dogs, spilled on ground, discarded
Total:	190 pounds 14 ounces	

The approximate total meat consumed in Tasbapauni during a 12-month period is presented in Table 21. Almost the entire amount (92%) was obtained from hunting and fishing. These totals do not include shrimp and green turtles[19] sold outside of the village, animals used to bait deadfall traps, nor spotted cats and river otters taken for skins. Of the total 110,600 pounds of meat, 6244 pounds (5%) were obtained by trade or purchase from outside the village;[20] domesticated animals butchered in Tasbapauni contributed 2490 pounds (2%); 101,866 pounds came from hunting and fishing efforts of the men of Tasbapauni.

The most important food animals are green turtle, white-lipped peccary, fish, and white-tailed deer. The high percentage of green turtle

19. Approximately 40,000 pounds of fresh shrimp and 82 turtles were sold. These totals are not included in Table 21.
20. There are two sources of meat traded and sold to villagers from the outside: shrimp boats and adjacent villages. Shrimp boat crews trade green turtles caught in drag nets and miscellaneous trash fish for provisions. The acquisition of meat from the shrimpmen is offset by dragging for shrimp close to shore in the "mudset" area where the Miskito like to set turtle nets during the dry season. Turtlemen in all coastal villages have lost nets to the shrimp boats, receiving no compensation. In nearby villages hunters and turtlemen who have made an especially large kill and cannot sell all in their own villages, may bring fresh and salted meat to Tasbapauni.

TABLE 20

Average Live and Butchered Weights (in pounds) of Important Game Animals, Tasbapauni.

	Butchered weight	Live weight
Tapir (*Tapirella* sp.)	250	525
Manatee (*Trichechus manatus*)	200[a]	500
Green turtle (*Chelonia mydas*)	90–100	190–210
White-tailed deer (*Odocoileus virginiana*)	60–65	85–90
White-lipped peccary (*Tayassu pecari*)	50	73[b]
Hawksbill turtle (*Eretmochelys imbricata*)	40–50	105
Brocket deer (*Mazama americana*)	40	60[b]
Collared peccary (*Pecari tajacu*)	40	55[b]
Paca (*Cuniculus paca*)	15	21
Spider monkey (*Ateles geoffroyi*)	12	20
Armadillo (*Dasypus novemcinctus*)	7	13
Agouti (*Dasyprocta punctata*)	6	9
Iguana (*Iguana* sp.)	6	10
White-face monkey (*Cebus capucinus*)	5	8
Hicatee fresh water turtle (*Pseudemys* sp.)	5[c]	15

[a]Plus oil made from fat.
[b]Said to be large.
[c]Including immature eggs.

TABLE 21

Pounds of Butchered Fish, Game, and Domesticated Animals Consumed in Tasbapauni, October 1968 through September 1969.

	Percentage	Pounds
Green turtle	70	76,860
White-lipped peccary	7	7245
Fish	6	7100
White-tailed deer	5	5800
Shrimp	3	3870
Collared peccary, brocket deer, iguana, manatee, tapir, paca, agouti, monkeys, armadillo, birds, shellfish, coati	3	2800
Pigs, goats, cattle, fowl	2	2490
Hicatee turtle	2	2380
Hawksbill turtle	2	2055
Total	100	110,600 pounds

in the Tasbapauni diet indicates how important this animal is to the villagers. The coastal Miskito have adapted much of their technology, lifeways, and internal and external economic patterns to predictable behavior patterns and relatively dependable catches of green turtle. By focusing on the green turtle a great deal of pressure has been taken off terrestrial animal populations which under other conditions would receive more hunting attention. Along the Río Coco, in contrast to Tasbapauni, there is less game meat in the diet and river fishing is the dominant source of animal protein.

The preponderance of green turtles in hunting and fishing returns and in the village's subsistence system can be shown graphically by depicting the numbers of the most important species taken monthly (Fig. 42). The 819 green turtles taken represent 819 separate successful instances of the culmination of skill, technology, timing, and effort.[21] There is no way of knowing how many turtles were missed, but the 819 are impressive by themselves. In addition, 133 white-lipped peccaries and 95 white-tailed deer were taken. Most deer kills took place during wet periods and the least during the long and short dry seasons. Except

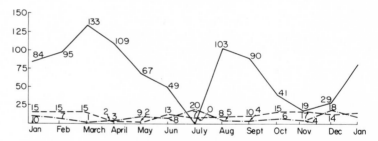

Fig. 42. Number of most important animals taken, Tasbapauni, October 1968 through September 1969.

(——):Green turtle	819	Iguana	40
(‐‐‐):White-lipped peccary	133	Collared peccary	2
(‐·‐·):White-tailed deer	95	Tapir	1
Hawksbill turtle	41	Manatee	1
Hicatee turtle	434	Brocket deer	1

21. Village turtlemen felt that this number of turtles was similar to the amount taken the year previously although no one ever bothers to count the total. They say, however, that only 10 years ago they got many more. Just how much is "many more" is impossible to tell.

One man who does count turtles is Weddy Ebanks, a Corn Islander of Cayman ancestry, who sets turtle nets off King's Cay during part of the year. He said that in 1967 he got 300 green turtles and in 1968 only about 225, using the same number of nets (32) and setting for approximately the same period of time. This man catches more turtles than any other single man on the coast. Except for an occasional turtle butchered for meat at his camp on King's Cay, all his turtles are sold in Corn Island or Bluefields.

for April and May when plantation work attracts the men, monthly totals of white-lipped peccary are fairly constant, reflecting to some degree the wide-range of this animal and the intensity of hunting effort directed toward it. White-tailed deer, on the other hand, are more localized, encountered less frequently, and in fewer numbers than are the wide-ranging droves of white-lipped peccary.

The low number of kills of animals other than those cited can be explained by the fact that very little attention is focused on them because they are either (1) thought to be more preferable at only certain times of the year (iguana and hicatee turtle), (2) occur on the margins of the effective range of the Tasbapauni Miskito (brocket deer, tapir), (3) require extreme patience and low return per trip (manatee), or (4) are considered taboo as a food (collared peccary).[22]

Seasonality of Meat Yields. Wild animal populations are not regularly available in place, quantity, or season. A close relationship exists between weather and sea conditions, local and long distance faunal movement, and the seasonality of meat yields in hunting and fishing. In Fig. 43 the line indicating monthly rainfall totals[23] also reflects other environmental and ecological changes. Periods of heavy or light rainfall are accompanied by differences in the velocity of offshore currents, changes in wind directions and intensities, and rearrangements of fish and game populations, most markedly the migration of green turtles. Meat consumed in the village reaches its highest volume during the dry season months of February, March, and April when weather and turtle conditions are optimum. A drastic difference in meat totals occurs in mid-May and intensifies through July and August as the turtles migrate southward and the rains begin. Hunting and fishing meat yields increase in the latter part of August and September during the short dry period. From September through the end of November, total monthly amounts of meat in Tasbapauni decrease as a result of labor demands for the rice harvest, contrary winds in October, and strong north winds in November. The absence of a morning land breeze, the presence of strong littoral flood currents, and contrary winds often prevent the Miskito from going to sea after turtle.

22. The two collared peccaries brought into the village were eaten by mixed Creole–Miskito families, not subscribing to the *kangbaiya* food taboo. Collared peccaries are frequently killed in the plantations as they root up the plants but they are usually left or dragged off into the bush. Hawksbill turtle, another *kangbaiya* animal, is taken for its shell. Some people will eat it regardless of the taboo, the hawksbill apparently receiving some consideration as a kin to the green turtle. The meat is always given away, not sold.

23. As explained previously, the only accurate precipitation records available for the immediate study are from Bluefields, 40 miles south of Tasbapauni. There is little difference in precipitation patterns between the two areas other than slightly lower annual and monthly totals in Tasbapauni.

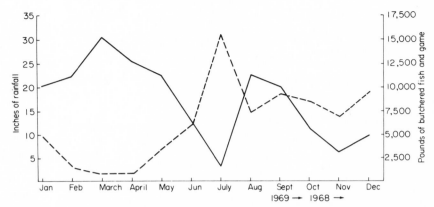

Fig. 43. Seasonal variability of meat yields from hunting and fishing, Tasbapauni, 1968–1969. Monthly totals of butchered fish and game (——) (in pounds) and rainfall (- - -) (in inches; 150.59 total; recordings are for Bluefield, 40 miles south of Tasbapauni) are shown. Amounts of rainfall also reflect the strength and intensity of currents and winds.

The variability of meat yields points out one of the most important ecological relationships between the Miskito and their environment, that of the interaction of weather and sea conditions, turtle migration patterns, and amphibian life of the turtlemen. The timing and success of turtling is largely dependent on these two major environmental factors. Turtlemen act as the adjustive link between the village's desire for meat and the vagaries and certainties of the sea and turtles.

Meat Yields of Biotopes. An attempt was made to measure the amount of meat taken from the different biotopes utilized by the Miskito. Meat totals presented in Table 22 represent meat taken by hunting and fishing for consumption within the village. However, many types of animals were taken by the Miskito only for skins. An estimated 40,000 pounds of shrimp were caught by Tasbapauni fishermen, but only 3000 pounds were kept for food, the remainder being sold. Only the most important species are shown in the table.[24]

The Miskito recognize many biotopes, mostly in terms of structural composition. They perceive the relation of specific animal species to certain biotopes and direct their meat-getting activities accordingly. Some biotopes were more productive than others because (1) they contained the desired animal species, (2) they were accessible, and (3) hunting and fishing were dependable.

Within the shallow offshore water environment, the most productive areas were the shoals, banks, and mudset biotopes, yielding 79% of

24. Many of these biotopes yield additional foodstuffs obtained from gathering and agriculture.

	White-lipped peccary	White-tailed deer	Green turtle	Hawksbill	Iguana	Fish	Shrimp	Hicatee	Miscellaneous (pounds)	Total
Ocean										
Shoals ("sleeping grounds")			27,150	1100		300			Whelks, Conch	28,550
Banks (feeding grounds)			43,205	855					Lobster[a]	44,060
"Mudset"			6,250	50						7150
Surf zone						450	400		*Ahi* 700	1150
Intertidal						1150				1150
Cays									Birds, bird eggs[a]	700
Land										
Beach		120			201					321
Old plantations	800	820								1620
New plantations	150	1600							Collared peccary 40	1790
Ridge and swamp	50	250								300
Marsh		300								300
Palm swamp	3050	1080								4130
Scrub thicket	100									100
Secondary	2000	600							Collared peccary 40	2640
Mature rain forest	300								Brocket deer 40	340
Gallery forest	180	360			56					596
Bankside ("vega")		520			44				Tapir 200	764
Freshwater										
Lagoon						400				400
Creek mouths						300	400			700
Shallow borders						600	350	800		1750
"Banks"						700	1850		Manatee 200	2750
Sandbars								105	Caiman, iguana, Crocodile, hicatee eggs[a]	105
Creeks						200		95		295
Grass flats								1355		1355
Totals	6630	5650	76,605	2005	301	4100	3000	2355	1220	101,866

[a]Not measured.

the total annual meat catch. On land, the palm swamps, old and new plantations, and secondary forests had the highest return of meat. In the lagoon–river resource sphere, the shallow water shrimp banks, grass flats, and shallow water borders provided most of the meat. In all approximately 87% of the meat was obtained from the water, and 13% from the land. Biotopes within the land and water zones are exploited in terms of cultural meat preferences, recognition of various species' habits and habitats, available technology, and the degree of subsistence risk.

The Tasbapauni population exploits less of their terrestrial environment than do the riverine Miskito. Hunting and fishing along the Río Coco are focused much more on small game animals (pacas, agoutis, brocket deer, monkeys, parrots and other birds, with only occasional white-lipped peccary taken). In hunting and fishing each kill is composed of a complex set of techno-environmental relationships. If the hunter or fisherman can base his quest on large animals, then each stalk, each shot, each trip has more of a chance for a large amount of meat. Therefore, a Miskito on the Río Coco may have to make five or ten separate successful stalks and shots for every one or two by a Tasbapauni hunter or turtleman.

Distances in Hunting and Fishing. In compiling data on hunting and fishing yields, round-trip distances to the "kill site"[25] were noted. Since 87% of the meat was taken from or transported by a *dori*, the round-trip distances shown in Table 23 are mostly direct water distances. Traveling by *dori* can involve little energy output if sailing is possible; however, paddling against a current can be extremely hard work.[26]

In Table 23 and Fig. 44, meat yields relative to distance are depicted. Each distance interval represents one in a series of concentric circles increasing outward from the village. Meat yields are shown as being equal at any point on a circle. A line passing from the village (at 0 miles) through all circles is shown in Fig. 44. Meat yields relative to distance are not on an even ascending or descending gradient. Instead the pattern is one of alternating highs and lows over distance. Thus it appears that meat yields from hunting and fishing do not closely reflect increasing labor or transportation "costs" from the point of origin of all inputs (the village) as one might expect from von Thünen's analysis

25. "Kill site" will be used to indicate the location where a particular animal was taken. Green turtles are not immediately killed, but are taken back to the village and kept until they are needed.
26. The Miskito have tremendous endurance paddling. Returning from a lagoon plantation in a large *dori* manned by three paddlers and heavily loaded down with provisions, I counted a *steady* paddle rate of 36 strokes per minute for 2½ hours.

TABLE 23

Meat Yields (in pounds of Butchered Meat) Relative to Round-Trip Distance from Tasbapauni, October 1968 through September 1969

Round trip in miles	White-lipped peccary	White-tailed deer	Green turtle	Hawksbill	Iguana	Fish	Shrimp	Hicatee	Miscellaneous (pounds)	Total	Percentage
0–4		120	95		35	1500	400		700 *Ahi*	2850	2.7
5–8	450	500			104	300				904	.8
9–12	1100	880	4,275		15	800			40 Collared peccary	6,460	6.3
13–16		1050	2280	50	16		100			4,596	4.4
17–20		820	2,660	100	13		250	45		3,928	3.8
21–24	1080	760	3,230	200	46	400	400	125	40 Collared peccary	6,281	6.1
25–28	1000	300	14,095	350		200	1200	775	40 Brocket deer	18,120	18.0
29–32	300	120	20,330	550	24	650	400	50	200 Manatee	22,424	22.3
33–36	700	200	11,780	355	30	250	250	55		13,620	13.5
37–40			5,795	100	18					6,253	6.1
41–44	1250	180	2,945					140	200 Tapir	5,365	5.2
45–48	750	720	1,900	100				990		3,645	3.5
49–52			7,220	200				175		7,420	7.3
Totals	6630	5650	76,605[a]	2005	301	4100	3000	2355	1,200	101,866	100.0

[a] Approximately 7660 pounds of this sold.

171

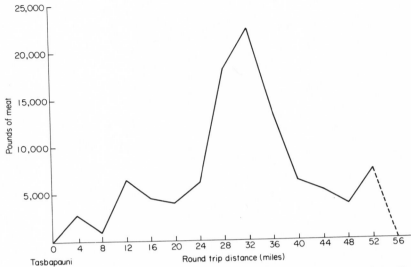

Fig. 44. Diagram of meat yields (in pounds) relative to round-trip distance from Tasbapauni, October 1968 through September 1969.

of agricultural location and yields as a factor of distance.[27] Actually the reverse might be true, that hunting and fishing yields may increase away from the settlement site because of greater isolation from human pressure. In addition, faunal populations, as we have seen, are not distributed evenly throughout an area. Instead, they commonly occur in restricted areas, under specific ecological conditions, and at certain times of the year (Nietschmann 1972).

Portraying hunting and fishing yields over distance without taking into account obvious environmental differences such as the distribution of land and water bodies, biotopes, and means of travel severely limits analysis. If we look only at the meat yields from the sea as a basis for distance considerations, some of these limitations can be avoided. Most of the village's meat comes from the sea, travel is by *dori*, and it is assumed that all sea conditions are constant (although this is not exactly true). In Fig. 45 meat yields from fishing and turtling at sea are shown. The highs and lows of the meat yields shown by the graph line from "0" (the village) closely represent the relative location and spacing of marine biotopes. Thus, going eastward from the village, the surf zone and close inshore waters yeilded almost 2000 pounds of fish and shrimp;

27. Von Thünen analysis is based on the assumption that certain activities will tend to fall in ordered circles around a central point (a city) because of increasing economic costs over distance (Hall 1966).

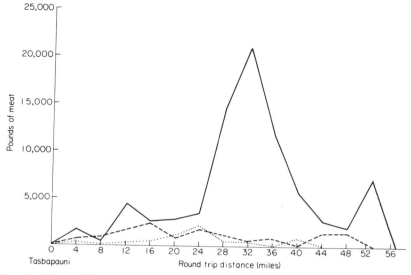

Fig. 45. Meat yields relative to round-trip distance by sea (——), land (---), and lagoons-rivers (····) from Tasbapauni, October 1968 through September 1969.

after this a decline in meat yields coincides with the fauna-poor longshore current zone; next, the meat yields rise at the mudset biotope where turtles are taken by net and harpoon during the summer dry period. Still moving seaward a deep-water zone of relatively low productivity is encountered. This is depicted along with the first turtle banks, thereby explaining why the graph line does not drop sharply. A rapid increase in meat yields is experienced at the distance intervals where most of the cays and turtle banks occur. Meat yields decline away from the first set of banks, shoals, and cays, over open water until the next cays, shoals, and banks are met, and yields rise again. For comparative purposes meat yields from the land and lagoon–river sectors are also shown in Fig. 45.

The long distances journeyed for hunting and fishing by the Miskito of Tasbapauni are similar to the situations in other coastal villages. The turtlemen in the Sandy Bay communities of the northern coast go 50–80 miles round trip to turtling grounds and 10–15 miles up the Ulang River for hunting. Just north of Tasbapauni in the villages of Río Grande and Little Sandy Bay, turtlemen travel 12–15 miles out to Man O' War Cays and 10–20 miles up the Río Grande to hunt. It is not uncommon for some coastal Miskito to have to go 1 and 2 days by dugout to upriver hunting areas.

Time Inputs. The expenditure of time spent on hunting and fishing provides a concrete measure for comparing their productivity and ascertaining the reasons for the patterning and scheduling of these activities. Estimating the amount of time involved in hunting and fishing was extremely difficult. Time inputs and meat returns varied tremendously from individual to individual and from season to season, much more so than in agriculture.

Not all turtlemen or hunters spend the same amount of time after meat, nor are they equally successful. In Table 24, inputs of time and meat yield outputs are examined for a hunter and a turtleman. These data are not sample averages but are drawn from two individuals, each known as being capable in his respective activity. Both of these men obtained approximately the same amount of meat during the initial 12-month study period. Each split the total meat take with a partner. The turtleman made fewer trips, but each of longer duration, than did the hunter. Both spent a great deal of time in travel: the turtleman 241 hours (53%) and the hunter 305 hours (57%). The turtleman had a slightly higher return of pounds of meat for every hour invested (1.78 versus 1.64) and considerably more grams of protein per hour (173 versus 138) than did the hunter. On the other hand, hunting provided more calories per hour of time (1220 versus 962) than did turtling, due in large part to the high fat composition of white-lipped peccary. The Miskito's liking for fatty meat prompts hunting efforts toward white-

TABLE 24

Time and Yield Data for Hunting and Turtle Fishing, Tasbapauni, October 1968 through September 1969[a]

	Pounds of meat (share)	Total hours	Hours traveling	Hours hunting and fishing
Hunter	875	533	305	228
Turtleman	812	455	241	214
	Pounds meat:hour	Calories	Protein (grams)	Fat (grams)
Hunter	1.64:1	677,044	73,317	41,496
Turtleman	1.78:1	437,500	78,750	7,875
	Calories per hour	Grams of protein per hour	Grams of fat per hour	
Hunter	1270	138	78	
Turtleman	962	173	17	
	Number of trips	Number of successful trips	Percentage of successful trips	
Hunter	26	14	54	
Turtleman	15	11	73	

[a]Composition of turtle, deer, and white-lipped peccary meats was adjusted from Wu Leung (1961) to allow for Miskito consumption of "mixed meat" and fat.

lipped peccary. This animal made up 70% of the total meat yield for the hunter in this example, while deer provided most of the balance.

The most significant difference, and the major reasons there are more turtlemen than hunters, is that the percentage of successful trips[28] was much higher for the turtlemen (73% versus 54%). Thus the turtleman went out from the village fewer times and had a greater chance of getting meat than did the hunter. In short, turtling reduces the subsistence risk of meat getting. It is not how much time or effort is expended to get a particular species that is important to a Miskito, but the degree of certainty in getting it.

Turtling has a lower subsistence risk than does hunting and supplies a higher protein return for time invested. Protein yield (grams of protein obtained per hour) for turtling is 20% higher than for hunting. Hunting appears to be more productive in calorie yields than turtling, but the margin of difference is insignificant because of the high agricultural productivity of calories. In addition, turtling seems to be a much more dependable (20% in this case) means of protein procurement than does hunting. Thus, Tasbapauni meat-getting strategy has reduced subsistence risk and increased protein availability by focusing on turtling (Nietschmann 1972). Even though the Tasbapauni people are situated along the edge of an extensive tropical rain forest environment, 65% of the active adult males concentrate their meat procurement effort on the sea and on turtles.

The timing of hunting and fishing is determined not only by environmental conditions, animal behavior, the scheduling of subsistence, but also by distance and day factors. Hunters make short expeditions for 1 day, 1 night or a day and a night at most. Turtlemen, on the other hand, may stay on the banks and cays from 1 to 6 days. The longer they stay out the greater their chances of success due to the increasing percentage of search time over travel time. Because of the long distances involved in getting from the village to the banks—from 15 to 26 miles—it behooves the men to stay as long as they can until they get one or more turtles.

The hunting and turtling week runs from Monday through Saturday; Sunday everyone is in the village for rest and church. Sunday as a day of rest is upheld by the three churches in Tasbapauni, and it is honored even by those Miskito who do not go to church in order to avoid social

28. A "successful trip" is somewhat difficult to define. A Miskito may bring a couple of hicatee turtle and believe he had bad luck hunting. I considered a hunting or turtling trip to be successful if the men brought back more calories in meat than they expended. In turtle fishing one either gets a turtle or he does not. Unless there is extreme meat hunger in the village and turtles are scarce, a Miskito will not strike a small turtle. If he gets one it will usually be 100–300 pounds live weight.

castigation by their peers. The necessity of being home on Sundays, of course, does not conform to turtle behavior. Turtles act much the same on Sundays as they do on Mondays. The "one should be home on Sunday" rule is a major pivotal point around which turtling and hunting activities must function. This would not be so important if it were not for the fact that hunting and turtling grounds are so far, and so much time is involved in getting there. If you do not have a turtle by Saturday morning, you must return to the village anyway and try again next week. In addition to regulating meat-getting patterns, being in the village on Sunday removes, at least for one-seventh of the time, exploitation pressure on game populations, as well as disrupting food-getting patterns. Starting from Monday then, until Saturday, increasing numbers of turtles are brought back daily to Tasbapauni as the result of turtlemen spending proportionately more time after turtle. Figure 46 depicts the per day percentage of one year's total of turtles brought into the village.

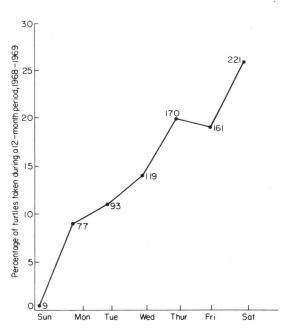

Fig. 46. Percentages of turtles brought into the village of Tasbapauni (green turtle and hawksbill) by days of the week, October 1968 through September 1969.

Factors of Exploitation

The forces of population growth and increasing dependency on out-side cash markets are intensifying and more narrowly focusing hunting and fishing efforts of the Tasbapauni Miskito. Furthermore, fauna within the Tasbapauni ecosystem are sometimes exploited by other human populations. For example, the heavy commercial shrimp boat activity, which occurs close to the shore during the dry season, inadvertently nets and drowns green turtles and hawksbill and prevents the Miksito from setting turtle nets in the mudset zone. In addition, continued exploi-tation of migrating green turtles off the Tortuguero Beach in Costa Rica is acting seriously to reduce the Nicaraguan turtle population, which, according to Carr (1972:29), contributes at least two-thirds of the Tor-tuguero nesting colony. Thus, more people, taking more faunal resources per capita for market as well as for subsistence, and continuing exploita-tion by non-Miskito populations, are combining to reduce the numbers and distribution of local wild faunas.

Several species of terrestrial and marine animals are in serious survival trouble because of overexploitation, prompted for the most part by world market demands for wild animal products. The focus of Miskito hunting and fishing activities is rapidly changing from subsistence-directed to market-controlled. The exploitation of animal resources in Tasbapauni is highly concentrated on specific culturally preferred food animals to satisfy subsistence meat requirements and village cash sale, as well as widely ranging to include other species sought for foreign markets. This is one of the major changes in hunting and fishing motivation in Tasba-pauni that we noticed in 1971, 2 years after our earlier work. The result is that populations of green turtles, white-lipped peccary, and white-tailed deer are attracting additional pressure from the Tasbapauni Miskito because of their taste preference and marketable potential, while at the same time several other species are being subjected to intensified exploita-tion. Hawksbill turtles, crocodiles, caimans, freshwater otters, jaguars, ocelots, margays, and lobster are all extensively depleted or almost exter-minated over much of eastern Nicaragua (Nietschmann 1971a, 1971b). As we shall see, large-scale faunal depletion serves to create ecological disruptions and generates additional social and scheduling pressures on the Miskito population.

The markets for spotted cat skins (jaguar, ocelot, and margay) have attracted more and more Miskito to hunt and trap these animals where before they were generally ignored save for an occasional kill for outside sale. The market for spotted cat skins expanded in the 1960s, by 1969 forcing the local price of a prime skin as high as $100.00 to $125.00

for a jaguar, $100.00 for an ocelot, and $7.00 for a margay. As market price increased in the 1960s, so too did exploitation pressure (Table 25).[29] After 1967–1968, the number of jaguars and ocelots taken declined while hunting focus shifted to margays.

The decline in the jaguar population, one of the major large predators, has led to an increase in white-tailed deer, and villagers in Tasbapauni are complaining more and more of resultant crop losses. And, with fewer remaining jaguars, hunters are less successful trapping and shooting the big cats. In the end the Miskito are the losers (along with the cats) for they frequently spend less time on their plantations or on turtling where yield potential is greater, and are finding greater crop losses. Furthermore, fresh meat is needed every other day or so to bait the deadfall traps used to catch the cats. Two men will usually have 10–20 traps, sometimes more. Multiplied by five trap lines operated by Tasbapauni villagers in 1968 equals from 50 to 100 traps. Monkeys and birds are favorite bait animals and are therefore rare in the areas where traps are being set, most having been killed or taken flight.

The size of Nicaraguan green turtle populations has been greatly reduced by several hundred years of export turtle fishing: first to victual visiting buccaneer and trade ships in the seventeenth and eighteenth centuries, then by visiting Cayman Island boats in the nineteenth and twentieth centuries for market sale, and recently by two new turtle companies established in 1969 and 1970 on the east coast. In the mid-1960s Cayman Islanders who wanted to turtle in Nicaraguan water were discouraged by the Nicaraguan government and since 1967 have been refused permission. The numbers of turtles taken from the waters off eastern Nicaragua have increased steadily since the two turtle meat companies began operations in Puerto Cabezas and Bluefields (Table 26). As one Miskito put it: "If de company wouldn't buy, dat amount of turtle wouldn't be catchin'."

Peter Matthiessen (1967), who accompanied one of the last Cayman turtling expeditions to the Miskito Cays, echoed the sadness for the passing of the turtles and the Cayman fleet:

> Standing at the hatch of the Adams, I peered down at the creatures. The turtles lay in vertical tiers, rack upon rack of broad, pale bellies extending aft into the gloom. Over the hold was rigged a canvas chute that caught the wind and funnelled it down for the refreshment of the cargo. "See dot?" Leewell spoke quietly at my shoulder. "Dey already lost dem pretty sea colors [p. 163]."

29. The legally recorded exportation of skins make up but a part of the total skins taken. Many more are not recorded in order to avoid taxation. Some skins leave illegally through Honduras, others via Costa Rica. Most skins are exported to the United States and Germany. A definite decline in ocelots is indicated by these figures and is borne out by Miskito hunters. Recent changes in U.S. laws, concerning the importation of skins of endangered species, should cause a decline in hunting pressure as local prices paid for skins drop.

TABLE 25

Reported Number of Spotted Cat Skins Exported from Nicaragua, 1966–1971[a,b]

Year	Jaguar	Ocelot	Margay	Total
1966	154	6123	285	6562
1967	231	5209	1677	7117
1968	466	2981	5440	8887
1969	354	1366	4034	5754
1970	246	1328	5237	6811
1971 (est.)	300	150	4800	5250
Total:	1751	17,157	21,473	40,381

[a]The majority of exported skins come from the east coast.
[b]Data from the Dirección de Recursos Naturales, Renovables, Managua.

TABLE 26

Estimated Number of Green Turtles Exported from Nicaragua, and Number Taken by Coastal Miskito[a,b]

Year	Number exported	Taken by Miskito
1958	1911	—
1959	2242	—
1960	1901	—
1961	1673	—
1962	2359	—
1963	2300	—
1964	957	—
1965	1218	—
1966	1148	—
1967	1263	—
1968	—[c]	4000 est.
1969	1000 sold to companies	5000–6000
1970	5000 sold to companies	8000–10,000
1971 (7 months)	4000 sold to companies	6000

[a]Data for 1958 through 1967 apply to Cayman Island boats; data for 1968 through 1971 refer to two turtle companies.
[b]Based on data from Dirección de Recursos Naturales, Renovables, Managua, and from "Frescamar," Bluefields, and "Tortugas, S.A." in Puerto Cabezas.
[c]No data.

The decline of green turtles has imposed new strains on the Tasbapauni Miskito's highly focused adaptation to the sea and to turtles. Turtlemen have had to increase their search time and extend turtling activities beyond the most opportune periods, creating time-binds with subsistence activities on land. Turtlemen in Tasbapauni claim that 20 or 30 years ago they used to get 50 or more turtles per week, while 10–20 is closer to the present village average. Two turtlemen in a *dori* used to be able to get their "load" (*dori* capacity) of four turtles in 1 or 2 days at most; now if they get three in a week they consider themselves lucky.

Growing human populations and growing markets prompted Carr (1967), who has done the most work on green turtle conservation in the Caribbean, to predict a bleak future for green turtles:

> It seemed to me that effective protection of the nesting colony alone would save the species. I no longer believe this. It is now clear that people are so abundant, and the life cycle of a sea turtle is so complicated, that nobody really knows what he is doing to a population when he kills a turtle or takes the eggs from a nest. The capacity of people to consume and their ability to destroy are growing beyond the tolerance of the small populations in which sea turtles live [p. 230].

Similarly, the hawksbill turtle is in a very precarious survival position due to the resurgence of the tortoise-shell market (Parsons n.d.), and the increasing market demand for its calipee, leather, meat, and eggs. Already considered one of the world's most endangered reptiles (Carr 1969:74), the hawksbill is among the most sought after of any marine animals found off eastern Nicaragua. Refuge populations are being depleted through intensive exploitation of local nesting beaches for turtles and eggs, and through large-scale turtle fishing of resident populations. In Tasbapauni, for example, approximately 107 hawksbill were taken by turtlemen in the first six months of 1971, an increase of 400% over the 1969 total of 27 for the same length of time.[30] This and other examples, indicate that market opportunities are dictating more and more the direction of hunting and fishing, which in turn will influence productivity as focus extends beyond subsistence.

The overall impact of overly focused and overexploitation of land and sea fauna is just being felt in Tasbapauni as cash alternatives are lacking and as subsistence skills are extended into market supply. Several maladaptive trends, intensified by changes in hunting and fishing patterns, are being manifested: (1) the economic is increasingly being separated from social considerations, (2) patterns of subsistence production and distribution are being disrupted, (3) dietary and nutritional well-being are declining, (4) subsistence scheduling is out of phase, (5) local human systems are losing subsistence resources to outside systems, (6) the stability, diversity, and control of the subsistence system are being lost, and (7) several animal populations are being severely reduced in numbers. These maladaptive trends will be discussed at length in the next two chapters.

---

30. While he was cleaning some barnacles from pieces of hawksbill shell, an inquisitive and perceptive Miskito asked me: "what do dey make from de hawksbill shell?" "Oh, combs, bracelets, earrings—things like that," I answered. "So, *dats* why de hawksbill is in trouble. Because of dis tings, he have to be careful."

CHAPTER VIII

THE ECONOMICS OF PRODUCTION AND DISTRIBUTION

Some people work all year and don't have and some people have and don't make a move all year.

—A Tasbapauni Miskito

Miskito society and the mode of production are no longer entirely organized around subsistence provisioning for domestic groups through kinship networks due to adaptations to cash wage and resource influences from outside, chiefly Western, systems. Nevertheless, production for market exchange is greatly constrained by the enduring values and organization of the production for use subsistence system.

In traditional Miskito society, an "economy" did not exist apart from a social context: The economy was something social groups did. Domestic production was for use or use-value; households produced according to their own livelihood needs and social responsibilities. If mischance should befall a household, their needs were supported by the kinship web. As production was regulated by what was needed (biological and cultural needs), surplus beyond subsistence was limited to an insurance margin to reduce subsistence risk rather than to an unfettered quest for profits. The Miskito then, were "underproducers" in that they did not realize the productive capacities of their subsistence livelihood systems. Sahlins (1972), in discussing the organization of production in primitive societies, noted that

> "underproduction" is not necessarily inconsistent with a pristine "affluence." All the people's material wants might still be easily satisfied even though the economy is running below capacity. Indeed, the former is rather a condition of the latter: given the modest ideas of "satisfaction" locally prevailing, labor and resources need not be exploited to the full [p. 41].

If a society does not have to produce more than it can use, or cannot use more than it produces, there is no need for surplus beyond subsistence or beyond exchange for use. The domestic mode of production, according to Sahlins (1972),

> harbors an antisurplus principle. Geared to the production of livelihood, it is endowed with the tendency to come to a halt at that point. Hence if "surplus" is defined as output above the producer's requirements, the household system is not organized for it. Nothing within the structure of production for use pushes it to transcend itself [p. 86].

If there is nothing within societies organized around production for use to cause production to extend beyond per capita requirements, then, obviously, increases must be influenced by contact with other systems and other organizing principles.

The shift in economic orientation occurring in Miskito villages is the major factor prompting these groups to transcend subsistence production, causing maladaptive social and ecological adjustments. In this chapter the economics of production and distribution will be analyzed, paying particular attention to the interrelationships and playoffs between subsistence and market, between provisioning and surplus, and between abundance and scarcity.

FOOD DISTRIBUTION

Indian have kind mind. Give you food; don't study dat. Give it away and when its done, its done. Dats de way de Indian like it.

—A Tasbapauni Miskito, 1971

Food dealings are a delicate barometer, a ritual statement as it were, of social relations, and food is thus employed instrumentally as a starting, a sustaining, or a destroying mechanism of sociability.

—Marshall D. Sahlins, in "On the Sociology of Primitive Exchange" (p. 170)

Food is not just to get, it is to distribute. As was seen in the preceding chapter, subsistence procurement and productivity are closely tied to environmental conditions and represent ecological adjustments between a population's nutritional needs and perceived environmental resources. Food is the product of the subsistence system and is the medium for the fulfillment of biological, social, and cultural demands.

Miskito domestic groups are not self-sufficient even though most of what they produce is consumed within the nuclear family.[1] In Miskito society it is almost as important to have enough food to fulfill social obligations as it is to have enough food to maintain a caloric minimum. Budgeting food resources between biological and social needs constitutes a major focal point for production strategy, decision making, and social alliances. As Rosmary Firth (1943), in her study of the Malay of Kelantan, observed, "The basic problem is universal: not only to have enough to eat to keep alive but to satisfy the demands of personal tastes, religious rules and a multitude of social obligations, all as important to the life of the group as mere subsistence is to life of the organism [p. iii]."

Food production and consumption are not constant in amount or in occurrence, household to household. A family will "have" one day and not "have" the next. Miskito families eat the same foods in roughly the same relative proportions; however, quantities vary depending on individual skill and luck of hunters and turtlemen, accessiblility to plantations, household composition, and the number and extent of familial and social ties to other food getters. Daily variations are reduced by the distribution of food through diverse social relationships.

An admired Miskito behavioral trait is generosity. A Miskito would rather give food to others and be known as generous than have plenty of food in the house and be looked down on as stingy and unfeeling for the problems of others. For example, a Miskito explained to me that "when there is plenty, it will be given; when there is none, there is none to give." In her study of the Miskito of the Río Coco, Helms (1967) indicated that "the rationale behind the sharing of food is generosity and a show of concern for the well-being of the other family or individual [p. 214]." As a result of the present economic depression on the coast, money is very scarce in Tasbapauni and subsistence resources are beginning to be sold in order to obtain money. This causes social conflicts between the seller and members of the village who would ordinarily receive portions.[2]

Among other things, food sharing and generosity are social mechanisms to reduce friction in the community over food accumulation. A

1. Sahlins (1968:75), in discussing family production in tribal societies, remarked that "household production is not precisely described as 'production for use'; i.e., for direct consumption. Families may also produce for exchange, thus indirectly getting what they need. Still, it is 'what they need' that governs output, not how much profit they can make. The interest in exchange remains a consumer interest, not a capitalist one. Perhaps the best phrase is 'production for provisioning' [p. 75]."

2. See Helms (1967:217; 1971) for an account of a similar problem in Asang on the Río Coco.

Miskito does not practice rigid "cost accounting" of food distribution, tallying how much goes out and how much comes in; nor does he distinguish food sharing as an economic process, an investment which will have an equal or better future return. Food distribution is a means to express and keep open social ties, which in turn function as ancillary channels for food getting. The distribution system acts as a regulatory means for equality of consumption, at the same time allowing recognition and expression of social obligations and community and family cohesion.

The distribution of food and other material goods in traditional Miskito society is not a distinct "economic" activity. In fact, the economics of exchange are hardly recognized as such and are merged with social obligations and relationships. Food sharing operates through a complex web of social linkages through extended families and converging on individual nuclear households. Each social relationship between kin, fictive kin, and friends represents one strand of the web and a potential channel for food giving or food receipt. Social relationships are maintained by the giving of gifts on a reciprocal basis. Most of the gift giving involves food distribution.

Meat is the most commonly shared food. Plantation crops are also shared but on a much less formal and obligatory basis. Cassava, bananas, plantains, breadfruit, sugarcane, and other crops are circulated via the same social distribution mechanisms and often supplement or bolster food supplies in other households. Proper respect for kin and other social alliances, however, is shown by giving meat: turtle, deer, white-lipped peccary. Since everyone does not fish or hunt, meat procurement is unequal, whereas almost every family has a subsistence plantation. Because meat is such a necessary part of the villagers' dietary patterns, it must be shared with others. One cannot stockpile or gorge on meat while others are without.

The overall rule for Miskito food distribution to eligible recipients is simple; it is the rule of two haves: "if have, have to give." In traditional Miskito society, if someone asked for food or was kin and needed food, the rule was that you had to give. Eligibility, position, or priority in social relationships (kinship distance), determine if one is to get meat or not. Clearly, today there is never enough meat for everyone. Close kin should receive first; more distant kin receive when there is enough. In traditional subsistence exchange, there is no question whether or not meat should be distributed to parents and siblings. If there is sufficient quantity it is sent.

The food-giving transactions within a social context involve generalized reciprocity. This is not to be thought of as "an eye for an eye." There is no expectation for compensation—when the food gift will be returned, in what quantity, or what type of food. Some will get back more than they give and others will give but rarely receive. Inequalities in food procurement are reduced; those who do not have or who are unable to obtain food, such as the old and the sick, will receive. Thus, the most productive food producers, the best turtlemen and hunters, will be the biggest givers and usually the smallest receivers. Their distribution of food to reciprocal exchange recipients often exceeds the amount of food returned via the same social channels. For example, a turtleman may distribute 30–40% of his annual catch to relatives, receiving in turn an equivalent of only 20%. Yet a Miskito would not consider himself "in the hole." The amplitude of differential reciprocal exchanges varies for each nuclear household and for each food-giving relationship. A range of 20–50% distribution "deficit" (the amount distributed minus the amount returned) was observed for meat exchanges during 1 year for three families. Each family head was a turtleman or hunter and capable of securing a meat "surplus" for reciprocal distribution.[3] But no malice or hard feelings exist if one does not carry his weight. "If have, have to give." In the end a certain balance in the village was achieved; at least minimal food requirements were obtained, food surplus or excess was disposed of to others, and social obligations were met as best as possible with available food resources.

Meat and subsistence crops designated for sharing with kin are divided into portions and sent by a man's wife. Those usually receiving a portion include the wife's parents, especially her mother, her sisters, her mother's sisters; her husband's mother, sisters and maternal aunts. Grandparents also receive portions as well as the wife's and husband's brothers if there is enough. Distribution of meat depends on the amount of meat obtained, size of the meat portion to be given, and the number and priority of recipients. Meat portions are generally 2–3 pounds in weight, with 1 pound usually being the minimum respectable amount one can give. A Miskito wife does not simply divide the amount of meat in hand by the number of sharing obligations and send equal portions to everyone. Close kin, especially women related to the wife and women on the husband's side, receive larger portions than do individuals further down the priority scale. If, say, only 10 pounds of meat are to be distributed, kin beyond immediate relatives will not receive any meat.

3. Food distribution mechanisms, amounts, and exchange differences constitute a research problem alone. See Henry (1951) for a detailed study of food distribution among an aboriginal group in Argentina. Each transaction involving the exchange of food was recorded noting who gave, to whom, and how often. In this way, Henry was able to isolate who were major producers and givers and who were major receivers, minor producers, and givers. The study showed that subsistence efforts were unequal, distribution was unequal, and receipt was unequal. Sahlins (1965) presents a number of examples of types of reciprocal exchange of foodstuffs among primitive societies.

An example of meat distribution might help clarify the previous discussion. An old turtleman and his son caught and butchered a 185-pound female turtle, obtaining 95 pounds of meat. They sold 40 pounds in 2- to 4-pound portions to 14 people. Fifteen pounds were given to a second son, in exchange for the use of his *dori*. The remaining 40 pounds were divided equally between the old man and the son who helped catch the turtle. The son had a separate house and family of his own. The old man carried his 20 pounds of turtle meat back to his house and gave it to his wife. Approximately 12 pounds were sent out to close relatives in 2- and 3-pound portions and 2 pounds each to two friends who were not able to buy meat at the butchering. Two second cousins came "to look meat" but did not receive any. The remaining 4 pounds were eaten by the man and his wife in 2 days. The individuals in this example are selling and distributing foods within a socially delimited system. A fairly typical pattern of this type of exchange and reciprocity is presented in Fig. 47. Other people in the village, as we shall see, are beginning to operate by selling what should be socially distributed.

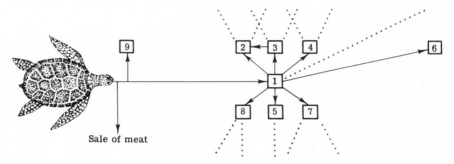

Fig. 47. Reciprocal food distribution linkages of Tasbapauni. 1: Primary producer; 2: Wife's mother/father; 3: Wife's sister; 4: Husband's mother; 5: Wife's brother; 6: *Lapia*; 7: Neighbor; 8: Godparent; 9: Turtle partner. (··········): other food distribution linkages not activated during this particular meat-giving instance.

Fictive kin relationships also require food sharing as an expression of mutual concern over the other's well-being. These are horizontal social relationships where goods and favors of equal kind are exchanged. One relationship, probably precontact in origin, is established when a person of the same sex is asked to cut the umbilical cord of a newborn child. That person is then called *lapia* by the parents and is considered family. This is the equivalent of being a godparent and implies social obligations and responsibilities between *lapia* and parent. In addition, the coming of the church has brought the idea of godparents, providing an added chance to cement relationships for parents. Often one birth will involve a *lapia* and godfather and godmother. Close friends who want to strengthen their relationship can become *libra* to each other by exchanging gifts and having a small party. People who are *libra* cannot fight or argue, and they always have to share whatever they have.

Other institutionalized food-sharing opportunities take place on Christmas and New Year's Day, when one is supposed to give food to kin, friends, and neighbors. In addition, a young man's first turtle will be butchered and distributed to older turtlemen as a mark of respect and to acknowledge that the individual is now a man and capable of bringing in and sharing meat.

Food sharing is also important in exchange labor arrangements. When men and women help plant they must be fed by the host family. Giving food to fellow villagers for agricultural assistance expresses thanks and spreads out the problems of obtaining food during the intensive labor input periods of clearing and planting when "looking a piece of meat" or "digging a stick of cassava" can detract from starting a new plantation. Common fare to give to a work group consists of turtle meat, cassava, and other breadkind, *wabul*, coffee, and bleached white flour bread. The importance of having food to give to people helping through communal labor exchange, is highlighted by the fact that even when turtle meat is sold, a person needing "planting meat" should be taken care of first.

The exchange of a gift of food between two parties represents more than a simple transmission of an item from one kinsman to another. Gift giving has a symbolism apart from mutual concern for kin. To be able to give is to be lucky enough to give and the luck of the gift comes from giving. To give a gift of food to someone is to keep viable the luck of gift giving.[4] Contained within the gift or within the symbolism of gift giving is thought to be a power which applies to reciprocity. If the luck of the gift is not passed on or returned then the power may be lost. In traditional Miskito society one should not profit at the expense of others; the gift is accepted but the luck should be passed on with another gift. This theoretically should cause a continuous circulation of materials in a Miskito village. In short, being able to receive is being able to give and to continue to maintain luck and equality. Receipt of a gift creates a feeling of discomfort to a Miskito which triggers a response in kind, but not necessarily in value. Thus in Miskito society, as in many other traditional societies, what is symbolically contained

---

4. The Miskito's description of the luck of the gift is similar to the description of Maori economics presented by M. Mauss in *The Gift* (see Sahlins 1972).

within a gift may be the reason for giving. To receive a gift carries with it a social and symbolic obligation for reciprocal giving in kind (see Sahlins 1972).

An example may help to illustrate the symbolism of the luck of the gift. If a person brings to you or sends to you a piece of meat on a plate, you must not wash the plate but should return it as you received it. If you were to wash the plate you would "wash the luck of the gift away" for you would be communicating to the person who gave you the meat that you "won't keep that person in mind for a return gift." The dirty plate is an acknowledgment of a promise to return a gift and to keep the power of the gift.

The Miskito say that one should keep in mind who gave them gifts in order to be able to return something when they are fortunate enough to have. It is the responsibility of the Miskito women to do most of the sending, receiving, and remembering in gift exchange of foods. Not only does this place the principal burden on women of maintaining the luck of the gift (we may recall Bell's (1899:246) comment that no hunter could resist the taunts of the women and had to seek meat), it also places them within a complex interpersonal communication system. I asked one husband how his wife could keep track of all the gift exchanges occurring over a period of time without forgetting and he answered, "what else them have to do?"

PURCHASED FOODS AND THE MARKET ECONOMY

These times, if don't have no money, no eat. Got to find your money.

You see what beat us here is dat there is no work. When de turtle stop and de shrimp stop, what to do? Right now little hardish to find a shilling to buy some flour.
—Remarks made by two Tasbapauni women, 1971

The production or acquisition of resources destined for village, regional, national, and international markets plays an important role in Miskito provisioning. Following Marx's "simple circulation of commodities," as exemplified by the formula $C \rightarrow M \rightarrow C'$, Sahlins (1972:83) related how the production of commodities (C) for sale in the market in order to obtain money (M) for the purchase of other, specific commodities (C'), duplicated primitive economic use-values of production for provisioning by exchange for consumption. Thus, the Tasbapauni Miskito have amplified and extended subsistence production to create resources to exchange for foodstuffs; that is, to exchange resources from the local system for desired resources from other systems by way of the wherewithal, money. The relationship between the two interacting systems becomes disproportionate if the amount of resources (or energy, information, materials) is not offset in kind by inflow of other resources. The outflow of protein from one system in exchange for the inflow of carbohydrates or money is not an ecologically balanced exchange even though the exchange value may be "equal."

Economic transactions involving purchased foods and money earned in the market economy form an important part of the over-all village food system. Some individuals have chosen to rely heavily on gaining their livelihood in the market economy, while others participate periodically, only using goods and money to supplement the more traditional subsistence patterns.

Average per capita income in Tasbapauni from the sale of market goods is high compared to most other Miskito villages. Proximity to green turtle banks, the rain forest, the lagoon shrimp banks, and to Bluefields via a year-round inland water route, gives Tasbapauni an advantageous location with respect to resource and market accessibility. More isolated Miskito communities and those utilizing less productive biomes and biotopes, are not able to participate as regularly in market activities as do Tasbapauni and a few other Miskito settlements.

Many combinations of different means for acquiring money are possible in Tasbapauni (see Appendix C). By looking at how an average Miskito family earns money, how much time it takes, and what portion is designated for food, it is possible to arrive at some comparability in analysis between the subsistence system and the local wage–market system as well as to find out how much the market economy contributes to subsistence. This ignores, of course, many of the other uses of money.

The families selected for analysis in 1968–1969 were fairly typical. Household budgets were kept for a year for three sample families, each of which had seven household members. During a 12-month period, the families averaged 1590 *córdobas* ($226.60), from approximately 1126 hours of labor (Table 27). The sale of turtles and turtle by-products brought 705 *córdobas* ($100.70) or 44% of the annual income. The single largest item of income was the sale of turtle meat in the village, which supplied 26% of the total.[5] There is a saying in Tasbapauni that "all money revolves around turtle." And this is so, for approximately 40–50% of most adult male's income comes from turtle fishing (green turtle meat, calipee, hawksbill shell). The rest of the sample families' money was derived from the sale of goods destined for outside markets (primarily Bluefields).

---

5. These estimates were made before the two turtle companies began buying from the Miskito. Today outside market sale of turtles contributes an even greater amount to a family's cash income.

TABLE 27

Average Amount of Money Earned by Sample Families from October 1968 through September 1969 (One Year) and the Approximate Number of Hours It Took to Produce or Obtain Each Item. The Values Are in córdobas (C$) (1 córdoba = $0.14)

	Price (córdobas)	Hours to produce or obtain item
Turtle meat in the village		
(825 pounds at C$.50 pound)	412.50	495
Three live turtles to Bluefields (half share)	82.50	45
Hawksbill shell (half share)	110.00	50
Calipee (65 pounds) (half share)	65.00	50
Turtle tag from Tortuguero[a]	35.00	1
500 Coconuts (15 córdobas per 100)	75.00	85
Coconut oil to Bluefields	150.00	90
One pig to Bluefields	75.00	100
Shrimp (20 cans at 10 córdobas each)	200.00	80
Rice (1800 pounds at 60 córdobas/100 pounds		
minus two-thirds costs)	383.00	130
$226.60 = C$ 1590.00		1126

[a]Archie Carr, a zoologist at University of Florida, has a turtle-tagging program at the Tortuguero nesting beach in Costa Rica. He pays a $5.00 reward for the return of a tag. It was due to my presence in the village that the turtlemen in the sample families sent in their tags which they had been saving.

TABLE 28

Estimate of Weekly Purchases and Amounts of Calories for Sample Families, Tasbapauni, October 1968 through September 1969

	Weekly purchases (córdobas)		Calories		
			per pound	per week	per year
Sugar	3 pounds (at .90)	2.70	1812	5436	282,662
Baking powder		1.50	—		
Flour	7 pounds (at .80)	5.60	1648	11,536	595,872
Beans	2 pounds (at 1.00)	2.00	1522	3044	158,288
Rice	3 pounds (at .80)	2.40	1635	4905	255,060
Salt	.5 pound (at .75)	.40	—		
Coffee	8 ounces (at .25)	2.00	452[a]	226	11,752
Meat	3.5 pounds (at .50–.75)	2.00	590[b]	2360	122,720
	C$ 18.60 = $2.65				1,426,354

[a]Coffee is mixed with filler.

[b]Calorie estimate includes meat, organs, and fat. The total is made up of three-fourths turtle, one-fourth white-lipped peccary and deer.

An effort was made to find out what percentage of the families' diet came from subsistence efforts in the direct production of food and what percentage was purchased. Weekly purchases of store foods were recorded for these same families of seven. Expenditures for store foods amounted to an average of 18.60 *córdobas* per week, or 1017 *córdobas* ($142.00) for the year (Table 28). Using calories as a common denominator and ignoring food value for the moment, each of the foods was given a calorie total for the year. This total was found to be 31% of all calories consumed during the same year. Therefore, 69% of the total calories for these families were produced by the families themselves, or acquired through nonmonetary based exchange with kin and friends.[6] It is believed that this figure is relatively typical based on dietary analysis of other families, which showed that subsistence foods contributed 65–80% of the diet.

It is apparent that a significant portion of an adult's activities in the village and regional market economy (Bluefields) is oriented toward obtaining money for food. In the previous example, 63% of the money obtained by the sample families from the sale of market products was used for buying food in the village. Purchases were made at a small store and from meat-selling turtlemen and hunters.

A crude way of estimating comparatively the productivity of participation in the market economy can be done if one assumes two things: (1) that breakdown of foods into calories is a relevant measure, (2) that purchased food-buying patterns are consistent week to week in a household. For purposes of analysis each hour of time spent on cash activities was assigned an equal money return; therefore, 63% of the total income was derived from 63% of the total hours worked. Each of the 709 hours (63% of 1126 hours worked on market products) of labor in the sample families that went toward buying food yielded 2012 calories.[7] It is neces-

---

6. Using calories as a common denominator, a measure ("degree of subsistence") of the amount of food from a traditional subsistence system can be easily obtained if the total annual calories consumed and the amount of calories in purchased foods or in produced foods are known:

$$\frac{\text{produced calories}}{\substack{\text{total calories consumed} \\ \text{(purchased and produced)}}} = \text{degree of subsistence}$$

i.e., for one man for 1 year:

$$\frac{750,000}{1,000,000} = 75\% \text{ degree of subsistence}$$

7. Calorie yield per hour of labor:

$$\frac{C}{F/I \ \times \ H} = \text{calories per hour}$$

(Continued on next page)

sary to emphasize that the relative percentage of different foods purchased are believed to be fairly constant from week to week. Obviously, if all the family purchased was sugar, then the calories per hour rate would be much higher.

Market exchanges occur at the nuclear family level while reciprocal exchange of subsistence items occurs within the extended family and wider kin and friend network. Production for use is limited by the producer's customary requirements which are culturally regulated. Sahlins (1972) observed that production for use "is under no compulsion to proceed to the physical or gainful capacity, but inclined rather to break off for the time being when livelihood is assured for the time being. Production for use is discontinuous and irregular, and on the whole sparing of labor-power [p. 84]." For the Miskito, production for exchange is organized to create a "surplus" above subsistence, or to exchange nonsubsistence items. Observe that what the Miskito exchange are forest and sea resources with "high ecological costs": protein resources such as turtle meat, shrimp, and pigs and high calorie sources such as rice and coconut oil; and second-order and third-order predators (spotted cats) and second-order consumers (hawksbill turtles); whereas in return, they purchase relatively low energetic cost items: foods from domesticated plants (flour, sugar, coffee) and material items produced in industrialized countries. Thus, if the market items are transferred from the Miskito's system to an outside system the relationship is asymmetrical; and if subsistence items are sold within the system to other members of the system it creates a demand for money which has to be obtained from the outside systems.

The gradual differentiation and distinction of economic value apart from social context in Tasbapauni is working toward increasing reliance on market exchange for livelihood at the nuclear family level, at the expense of reciprocal exchange within the extended family. If purchased or produced for market, the new market behavior rules apply and the item will not be shared beyond the nuclear family. On the other hand, if it was produced for subsistence then it will be shared within a distribution system regulated by kinship ties. The friction results when subsistence items are removed from social channels in order to be used in

---

where C is the total calories in purchased food, F is the money spent on food, I is the total income, H are the hours worked

For a Tasbapauni family example:

$$\frac{1{,}426{,}354 \text{ calories}}{(1017 \text{ córdobas}/1590 \text{ córdobas}) \times 1126 \text{ hours}} = \frac{1{,}426{,}354}{709}$$

$$= 2012 \text{ calories per hour for adult man and woman}$$

market exchange. Thus the social value is taken from the product and given a market value instead.

THE TRANSITION FROM SUBSISTENCE TO MARKET

See, what de problem is, everyone wants to live off money. People following de money just like ants de sugar. De way de place going, everyone moving for demselves.
—A Tasbapauni Miskito, 1971

Market transactions involving food generally act in direct conflict with reciprocal exchange patterns of subsistence foods. As Sahlins (1965) observes for primitive societies: "Food has too much social value—ultimately because it has too much use value—to have exchange value [p. 173]." However, as a group's economic system is transformed from subsistence to market orientation, food begins to have exchange value within the community. For the Miskito, engagement with the market necessitates the accumulation of food and materials in quantities for selling, while at the same time somehow having enough to acknowledge traditional social relations and expectations. In Tasbapauni, a Miskito's livelihood strategy involves attempting to satisfy diametrically opposed goals with limited means.

In an isolated domestic unit of subsistence production (isolated household) the level of labor intensity theoretically would be closely geared to the number of mouths there were to feed proportional to the number of workers, or better said, to the ratio of consumers to workers.[8] However, in any given society, the intensity of production is not related solely to provision for household members. The social system of a group arranges domestic production levels and reciprocal distribution mechanisms to fulfill consumption demands of households which fall below minimal subsistence levels because of sickness, old age, crop loss, bad luck hunting and fishing, and so on.

8. The Russian economist, Chayanov (1966), observed that in subsistence production, "the volume of the family's activity depends entirely on the number of consumers and not on the number of workers [p. 78]," other things being equal. He proposed that it was the ratio of consumers to workers which set the intensity of household labor efforts (see Fig. 48).

Fig. 48.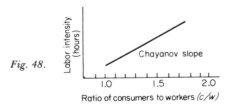

Ratio of consumers to workers *(c/w)*

In traditional Miskito society, each of the productive nuclear households is responsible to a wider group (kin, friends, neighbors) for food and general assistance, often putting a strain on an individual for food sharing beyond his immediate house. Food exchange among kinship groups tends to alleviate inequalities, but it still places the principal burden on a core of food producers around whom the subsistence system revolves. Within a subsistence-based society the nuclear family is not an isolate but an integral part of the social fabric of the particular community; reliance upon, and obligations to, kin and friends are paramount factors in the subsistence system and in the intensity of production.

For the Miskito, food is losing its social value as it takes on monetary value. If village economic organization continues to change toward monetary-based market transactions at the expense of reciprocal exchange, then its social complement will be the increased individualization of households, the fragmentation of community structure, and possible rearrangement or termination of subsistence economy-embedded social relationships. In the end, "the fortunate households cannot be responsible for the unfortunate [Sahlins 1965:180]."

With the advent of increased transformation to a market economy, reciprocal exchange patterns among villagers are being changed. Social relationships have been very elastic in their response to economic changes in the villages, permitting reduction in the amount and frequency of food sharing, as well as allowing satisfaction of kin obligations by a "go to the head of the line" policy in local economic transactions. The relationship of social linkages and reciprocal food exchange has acted as a buffer against declining meat yields and increased pressure toward social fragmentation through participation in an expanding market economy.

Purchases of flour, sugar, rice, coffee, and baking powder at one of the small stores in the village involve small amounts of food, enough for the day and no more (Appendix D). Money is obtained in spurts: the sale of some coconuts, deer meat, a few pounds of calipee. Needs for store bought food are constant, a necessity. Market exchanges are made in small amounts, such as a "shilling sugar,"[9] "a *córdoba* flour." Store goods are made available to most villagers on an even basis by extension of credit payable when the rice harvest is sold or when shrimp start running. Without credit, most Miskito in Tasbapauni would be unable to buy part of their daily food supply at the store.

---

9. A 25 centavo coin is known as a "shilling" on the east coast, 50 centavos as "fifty cents" or "two shillings." This is a carryover from when English and American currency was used on the coast.

Scarcity and Surplus

> *Everything dats money get scarce.*
> —Tasbapauni, 1971

For the Miskito, economic entanglement with larger, more complex systems has occurred sporadically and at different levels of intensity over several centuries. Boom-and-bust economic cycles have created markets for labor and resources and are transforming them into commodities with a price, a value. The Miskito are now confronted with problems of how to create a market surplus out of scarce resources.

The Miskito consider ecosystem resources as having a value and because they have a value, they are scarce. Shortlived economic opportunities for working with foreign companies or selling forest and sea resources have instigated a reciprocal reaction from the Miskito: "Anything good never lasts long so when it comes you better make good of it, like the hawskbill. Anything dats valuable get scarce in this country. The alligator, the tiger, all are scarce now," as one Tasbapauni villager expressed it. Therefore, if human and environmental resources have a value, and that value can only be realized for short periods, then labor availability of males and exploitation rates have to be adjusted for quick returns: "When company there, got to get. When dey leave, dats gone blank; dey no looking you." As discussed previously, absence of males from villages for company work, was assisted by subsistence and social adjustments by women: women's swiddens, operated by women in order to guarantee subsistence foods, and a tightening of social interaction between women in the village. In addition, cash market opportunities for forest and sea resources focused and intensified exploitation of specific valuable species. So as to free labor and resources for market, socially regulated subsistence production and distribution patterns had to be altered. Low-labor investments on seasonally, spatially, and species-dispersed exploitation for subsistence are changing to higher labor inputs on specific species and biotopes.

Because Miskito subsistence was organized as a low risk, low labor demanding system which could produce relatively high and dependable yields over a yearlong period, surplus production of food was limited. Surplus output beyond extended family or village needs was built into the agricultural system, for the Miskito planted more than they could eat as a hedge against possible food loss due to flooding or a hurricane. The diversity and accessibility of wild animals was great and one had but to spend a short time hunting and fishing in order to secure additional protein supplies.

When weighing scales were first introduced to Tasbapauni in the

1920s, they were used only for transactions involving sales to outside markets. During this period, Miskito villagers say that no one could sell subsistence foods in the village because "everyone had, there was no market for ground provisions or meat." As the use of the scale and cash value equivalent spread to locally used resources and blossomed into internal village sales, resistance to weighing and pricing was expressed. According to one informant: "First time de scale business not here. When came, Indian people said was spoiling de place. Now playing wit de scale; Indian don't like dat."

As participation with a market economy or production for exchange expanded from exchange with outside systems to exchange within the system, production and distribution patterns were affected. In order to create a "surplus" or subsistence foods for village and external sale, an individual or a nuclear family could: (1) increase production beyond subsistence, (2) reduce the amount and frequency of materials and labor reciprocally distributed, (3) reduce the amount of an item needed for subsistence. The first option requires either monetary outlay for the hiring of additional villagers to help with market-oriented cultivation or increased labor inputs from the nuclear family; i.e. more time spent in cash agriculture, hunting, and fishing. The second and third alternatives induce the individual or the nuclear family to cut down on either or both the quantity and occurrence of gift giving to kinsmen and friends or the levels of consumption within the family. Each of these alternatives might be considered a strategy for the creation of a market surplus in which the relationships between subsistence production intensity, ecological stability, social cohesiveness, and dietary well-being may be altered.

Turtles: Out of the Pot and into the Pocket

Turtle getting scarce, so will have to hold on to what is yours.
Turtle is de lifeline of de people.
—Tasbapauni, 1971

In the previous sections, some generalizations were made pertaining to the strategies and effects of market-oriented sales within a subsistence society. The following will place these observations in a specific context: the impact of outside and village market sale of turtles on the Miskito of Tasbapauni.

Turtle meat is never abundant enough to satisfy kin obligations, village market sale, and the commercial market. "All can't get because there is not enough. Dats what de problem is," said one Tasbapauni turtleman. Turtle availability is sporadic in occurrence according to environmental conditions, limited in supply, and in high demand. It would take about five medium-size (200 pound) green turtles to supply the daily demand for meat in the village. From April 15 through May 15, 1971, a good turtle period, more than 125 turtles were caught by Tasbapauni turtle-men, but only 34 were butchered in the village, or only about one turtle per day. From the meat supplied by the 34 turtles came portions for gift distribution to kin and turtle meat to sell to villagers.[10] The other 91 turtles were sold to the two commercial turtle companies.

An early morning turtle butchering is the most important recurring event for social interaction in Tasbapauni. Economic and social relation-ships are played out every morning in terms of to whom meat is sold, in what order, and in what amounts. Portions of meat taken from the butchering on the beach will be distributed by the turtlemen's wives via children to selected people; thereby visibly demonstrating social linkages.

Only the closest of kin can be assured they will receive meat from a butchered turtle. More distant kin and other villagers can buy meat, but again there is never enough. One would assume, in terms of Western economic organization, that the market economy is independent of the social system, but this is not so in Tasbapauni. The turtlemen and the hunters sell to whomever they choose, not the first in line. Therefore kinship and friendship relationships can be acknowledged by selling meat to those who would not ordinarily receive meat, being socially too distant.

Preferential selling of limited amounts of meat to some individuals over others, is referred to as "looking on faces." In order to ensure chances of obtaining meat adult men often go to the butcherings in place of their wives or children. Turtlemen are more apt to sell meat to other turtlemen or members of their family to ensure reciprocal treat-ment. Consequently, much of the meat that is sold circulates among the best turtlemen: "When I is butchering, I look for de turtlemen. Dey is going to get. When dey have, den I is going to get," said an informant.

At any one turtle butchering there may be 30 or 40 men, women, and children all yelling "sell me two pound," "sell me three pound" at the turtlemen as the meat is being cut up. The turtlemen seemingly pay no heed to the din about them; head bent down they chop meat

10. Prices for meat in Tasbapauni were 50 centavos ($.07) for turtle and 75 centavos for deer and white-lipped peccary. Hicatee, pig, beef, goat, manatee, and fish are also sold, but most other animal flesh is considered second class and unmarketable. Hawksbill meat is given away. Small game animals are never sold; sometimes they are presented as a gift, but usually they are eaten by whomever shot them, never achieving the penultimate status of real meat, worthy of a gift or for selling.

Fig. 49. Selling turtle meat in Tasbapauni.

into "junks" with machetes (Fig. 49). When they are finished butchering and some 100 pounds or more of turtle meat are piled before the crowd, the noise reaches a crescendo. Expressionless, the turtlemen make their sales, each transaction followed by a shout of joy and wide smiles by the recipient. The two partners, one butchering and weighing and the other taking money, must keep a running log of who should get meat and if the quantity asked for is sufficient or too much for that particular family. No one is specifically denied meat—that would be too much of an afront, but many will have a better chance because they are kin, or are turtlemen or hunters.

Meat portions are restricted to 2, 3, or 4 pounds at the most so as to increase the likelihood that people will "get." As the pile of meat grows smaller the cries become more pleading, more highly pitched. Still the turtlemen show no emotion, rarely speaking. They are acting out a split role: they have brought meat to the village and that is good, but they have to sell it to get needed money and therefore they have to chose who will get and who will not. The amount of meat sold depends somewhat on how much pressure is put on the turtlemen for meat. He must keep enough for his family and for giving to some relatives, but if meat has been especially scarce he will cut off a few pounds from his portion and sell it. Still there will be many who are unable to buy meat; some will curse the turtlemen who are butchering. Finally with a sad, hurt look, a last rattle of their bowl against the side of the *dori*, they walk away, without meat for the day. And without meat to eat, a Tasbapauni Miskito is simply without sustenance: "No meat, neither fish, dats what's beating me now," a Miskito remarked.

For those deprived of the chance to purchase meat and those who had no money that day even to attempt, it will be another meatless day, with only boiled cassava, *duswa*, and bark tea eaten in the quiet of the kitchen, the midday sun burning against the fire-blackened walls. Late in the day new hope for meat will come—a sail on the horizon, a *wari*-filled *dori* gliding through a back creek. Once more cries of "meat, meat" will be heard in the village.

The reason why some of the most skilled turtlemen in the world, exploiting one of the world's richest surviving turtle grounds can't satisfy the meat cravings of a village with but 1000 people is simply because time and turtles are being siphoned from subsistence to market. The cumulative results were evident during our 1971 fieldwork. The demand and clamor for meat was greatly increased at the turtle butcherings, some women even hitting the turtlemen on the head with their pots. Social tensions were noticeably more evident; people openly complained of being deprived of meat. And even though the turtlemen were spending more time at sea after turtles their returns were much lower per trip, per hour, than during our 1968–1969 fieldwork.

In late 1969 a foreign-owned turtle company began buying and freezing turtle meat in Bluefields for export and in 1970 another turtle company, supported in part by the Nicaraguan government, began operation in Puerto Cabezas. The companies send boats to Miskito villages and to the turtle camps on the offshore cays to purchase turtles at $8.60 each for all over 125 pounds. Demand is unlimited.

With a guaranteed market and high prices, the Tasbapauni Miskito, as well as other coastal Miskito communities, began to intensify exploitation of green turtles (Table 29). The numbers of turtles taken for the first 6 months of 1971 increased 228% over a comparable period in 1969. The sale of turtles outside of the village incrased 1500% while the amount of turtle meat consumed in the village decreased by 14%. Outside market sale of turtles in the first half of 1971 almost equaled the number of turtles that had supplied most of the village meat in 1968–1969.

TABLE 29

Comparative Totals of Green Turtles Taken and Sold by Tasbapauni Turtlemen, 1969 and 1971 (Recorded and Estimated)

	Butchered in village	Sold outside of village	Total
1969 (January through June)	437	47[a]	484
1971 (January through June)	374	731[b]	1105

[a]Sold in Bluefields.
[b]Sold primarily to turtle companies.

More turtles are being caught, more are being sold, but less meat is available for food in the village. Turtlemen are investing increased amounts of time to secure marketable protein for outside sale, and then purchasing largely carbohydrates from external sources with the money earned.

Some of the factors determining whether or not a turtle is to be sold to the company boat or butchered in Tasbapauni include: (1) the size of the turtle, (2) the amount of money to be earned, (3) commitment to social obligations, and (4) the demand for meat in Tasbapauni. A few turtlemen are beginning to emphasize monetary return over social responsibilities in deciding where and to whom they will sell. Money gained from butchering turtles in Tasbapauni comes from the sale of meat to the villagers and calipee to buyers for shipment to Bluefields. In general more money can be earned from turtles larger than 200 pounds by selling in the village rather than to the company (Table 30). The older and larger age classes of turtles have been reduced by overex-

TABLE 30

Differential Returns from Selling Turtles in Tasbapauni and to one of the Turtle Companies, 1971. Estimates Figured on the Basis of Average Weight of Butchered Meat and Calipee per Turtle

Live weight (pounds)	Tasbapauni (meat and calipee) (C$)	Company (live turtle) (C$)
125	37.00	60.00
150	46.00	60.00
175	52.00	60.00
200	61.00	60.00
250	75.00	60.00
300	88.00	60.00

ploitation so that large turtles are becoming more uncommon and harder to find. If the commercial demand for turtles exceeds the receipt, the companies might raise the amount paid to further encourage Miskito turtlemen. Therefore, the on paper economic breakoff point between sale to the company or in the village would be for even larger turtles.

As turtlemen sell more turtles to the company boats and bring fewer turtles back to Tasbapauni, pressures for meat in the village increase. For example, a turtleman who brought in three turtles after several meatless days was besieged from late afternoon to 11:00 PM that evening by people requesting and pleading with him to sell them a piece of meat the next morning. To avoid the unwelcome task of selling meat to a limited few when according to traditional Miskito behavior all should really receive, many turtlemen are selling to the companies and returning to the village with but one turtle for their own family and close kin. Whereas before, one was considered stingy if he sold meat instead of giving, at present he is thought to be miserly if he does not sell! Similar rules but different contexts. One of the best turtlemen in Tasbapauni, for example, has sold to the company 70 of the 80 turtles he caught between January 1970 and May 1971. Many other turtlemen will not sell meat to his wife because he never has meat to sell to them. A common remark made to individuals who have withdrawn from meat giving and selling in the village, but still want to buy meat from others is: "Let dem cook de money."

Greater dependency on turtle fishing for money, declining turtle populations, and the larger amount of time required to catch turtles are creating severe dislocations in subsistence scheduling and meat consumption patterns. A few years ago, according to informants, Miskito women followed the traditional practice of sun-drying 2 or 3 pounds of meat from every turtle their husbands caught, and saving it for the wet season meat famine months: "When de rain on and feel de heft of de weather, no have to go nowhere, nor grumble for meat. We just sit and watch what taking place and have something to eat." Now, however, few families store meat and if they do they are subject to requests for a piece of meat from kinsmen.

The older turtlemen in the village claim that the younger people are not "watching de distance." That is, they are not anticipating the distant consequences of participating in market transactions in the village: "You should carry de distance first and den refer back on de thing before you make your move. Meat is above de distance now. De younger race just made a mess of things and can't stand it." The older Miskito counsel against selling in the village; food should be given, it should be shared. But they note that "through de money, de young begin to give up sharing. Through dat selling business we'll lose de luck of de gift."

The more dependent Tasbapauni nuclear families become on turtles (as well as other marketable resources) for international external exchange, the more independent they are becoming from extended families and the kinship network. To become dependent on a declining resource is to enter an ecological blind alley where the exploitation market system is accelerating in intensity because of positive feedback generated from the disruption of the subsistence sector and its rules for social expectations and economic behavior. To the extent that families participate in cash market activities involving not only surplus resources and labor above subsistence, but also labor and resources *from* subsistence, is the degree to which they have to disengage from horizontal social relationships kept viable through reciprocity. A Miskito woman portrayed this relationship by observing, "Have to work. First time family give, not now, have to buy." Both the nuclear family labor supply and the biotic resources exploited for market are limited. To satisfy the need and demand for money, human inputs have to be individualized and increased within the ecosystem in order to divert production out of the system and to realign distribution in the system.

THE OLDER HEADS AND THE YOUNGER RACE

De younger race dem go in de night to kill a deer but dey is killing plantain and banana.
—An "older head" in Tasbapauni, 1971

Conflicts between stipulated social behavior and contradictory market behavior with land, labor, and biotic resources are becoming frequent and there are major disagreements between generations over the means and methods of livelihood. The elder Miskito, older than about 30 and referred to as the "older heads," often criticize younger adults ("younger race") for their economic orientation toward production for exchange and lack of reciprocal distribution of food gifts: "Pleasures of de money will be no help to you in what to eat."

Differences in behavior and values between the two groups cannot be easily broken down into generalized patterns as there is considerable overlap between old and young and between those said to have "Creole ways" and those with "Indian ways." The overall differences, however, are sufficient to influence subsistence production in the village.

The "older heads" have a background of working for companies during the banana, gold, and lumber boom periods prior to 1940. Subsistence behavior in the village was largely traditional, based on diverse resources exploited by communal labor and distributed through generalized reciprocity. The younger race represent a different era of economic history on the coast; a "bust" period, most of the companies had left, and money had to be obtained by exploiting local resources according to available outside market opportunities.

Discrepancies between the younger race and the older heads are reflected in the following comments toward subsistence and market recorded in Tasbapauni in 1971. The older Miskito maintain that everyone should plant breadfruit trees so that everyone will have some during periods of bad weather or food shortages. The younger Miskito say that if everyone plants breadfruit there will be no one to sell breadfruit to. As to swidden work, the elders advise that every family should plant enough for its needs and enough for reciprocity, while the young Miskito remark that "plantation work is not nourishing, just labor. No pay to work," or, more to the point, the work does not pay. Furthermore older heads say that one must not "study de turtle all de time." Instead one should "tend to plantation." On the other hand, many young Miskito feel that they "have to study de turtle all de time, because dat's where de money is." The older heads say that all the young men bother with are turtle and shrimp and that when these are gone all that will be left will be coconuts and rice: "All is money, dey no think about de family."

From our observations and field data, young adult Tasbapauni Miskito generally have smaller subsistence swiddens, tailored more to the nuclear family, but not as large as the swiddens of older Miskito. They do spend more of their time attempting to acquire resources for sale and their livelihood strategy is much more simplified and focused than their parents'.

The young men are frequently accused of stealing subsistence swidden crops for home consumption. This appears to be true. Many young men have diverted their labor inputs from subsistence agriculture to resource procurement for exchange which supplies only intermittent monetary return. Family food needs are constant and have to be satisfied by other means if money is not available and the subsistence swidden is too small or the crops are not ready for harvest. Complaints of stolen crops are common and many verified incidences have been seen. The older Miskito are skeptical when they see young men leaving the village at night to "torch deer" with a flashlight and rifle: "They no torchin' deer, they torchin' green skins."

Although many people say they know who steals food, few attempt to do much about it except get mad. The act of stealing bridges two economic systems and two ways of life. One steals because of the inability of the cash market to supply money for sufficient foods but one is protected from harsh treatment for stealing because of traditional social considerations. Since the Tasbapauni Miskito feel that everyone has a right to food and traditional exchange patterns are set up to supply food to those who fall below a subsistence minimum, then one should not be too "hard" on a fellow villager for taking something that only recently had a commodity value in the village.

A Tasbapauni Miskito woman who was concerned about thefts from her husband's swidden said that the reason the Miskito are Christians is so that they can go to church and pray: "Some praying dat people no steal their provisions. Others praying for rain so dat dey can steal."

Stealing has adversely affected agriculturalists by more than occasional crop loss alone. Even though a Miskito considers himself to be generous, stealing from others is said to be wrong for it demonstrates little concern for the well-being of the family who planted the swidden: "I no going to work for others when dey no care for me." Stealing appeared to have increased in the village since our earlier fieldwork. Judging from our 1971 observations, the reaction has been to plant smaller subsistence swiddens to curtail the degree of loss due to stealing. Consequently, the increasing frequency of stealing by a small proportion of the population has precipitated a decline in subsistence production efforts and, accordingly, the scale and range of subsistence food distribution in the village.

CHAPTER IX

FOOD PREPARATION
AND CONSUMPTION

When food come eat what you can—don't wait for tomorrow.
—A Tasbapauni Miskito, 1971

The Indian has an enormous appetite when there is plenty of food, especially fish and meat, at hand; on the other hand, in case of food shortage, they are able to travel and work with scarcely anything to eat.
—Eduard Conzemius (1932:88)

The preparation of food is a vital aspect of the food system, being the transfer point where raw materials are converted into edible foods suitable to Miskito dietary preferences, food needs, and storage and preservation techniques. Too often ecological studies treat human diets simply as the gross material to fuel a thermodynamic machine—so many calories in, so much expenditure of calories out. For the Miskito however, the types of foods and the ways in which they are prepared determine to a large extent whether or not one feels like doing hard work regardless of the caloric or nutritional content of the foods ingested. Miskito diet and nutrition are the end result of the entire food system, reflecting the system's efficiency and ecological adjustment. Dietary analyses can show how the subsistence system functions by indicating in a concrete manner what percentage and what quality of food production per person arrives on the table.

205

FOOD PREPARATION

The ways in which foodstuffs are prepared and stored are, in part, cultural adaptations to the ecology of the Miskito's habitat. Various food cravings and preferences are in many cases nutritional signals for a type of food needed by the body. Precise methods of preparing desired foodstuffs make them fit for Miskito consumption. Food preparation techniques are means of adjusting culturally and physically available foods to Miskito dietary preferences—the all important middle steps which link food resource with food appetite.

Although simple in techniques and using primarily local food resources, Miskito culinary practices yield highly palatable dishes. In short, Miskito food is good food. This is not just a personal endorsement for Miskito food, but an attempt to counter frequent statements on the supposedly bland and insipid aboriginal foods in Latin America. To be sure, some Miskito cooking methods have been influenced by contact with other groups, especially Creole peoples; however, the core remains Miskito.

Most food is cooked by boiling in water, roasting over a fire or in ashes, and, for a few foods, frying in coconut oil. Because cooking methods are not elaborate, kitchen utensils are simple and few. The focal point of food preparation efforts is the U-shaped clay hearth (*kubus*) supported 3 feet or so off the floor by a wooden frame (Fig. 50). Women maintain a perpetual small bed of hot coals, ready "to make fire" if meat should be brought in. Firewood is gathered from driftwood on the beach and from newly cut plantations.

The beach supplies a good deal of the kitchen utensils used in Tasbapauni.[1] Coconut oil used in cooking is stored in plastic bottles which are found on the beach. Large metal net floats are cut in half and make satisfactory cooking pots. Wooden implements are usually made from cedar or mahogany and every kitchen has an assortment of wooden spoons, shallow bowls, and beating sticks (Fig. 51). Large wooden mortars are used for hulling rice for home consumption and for pounding a variety of plant foods. Flattened out tin cans with nail holes punched through are used as coconut graters. Calabashes and tin cans serve as

---

1. The beach is a major resource-yielding zone for the coastal Miskito. Besides the bottles and floats (used as canteens on turtle expeditions), the *awra* (floatsam) may produce cut lumber for houses, sheets of plywood, hatch covers, logs, rope, nets, cork and polyester foam floats, pieces of rubber and plastic used for glues, and occasionally current bottles with a reward of $.50 for return of a postcard. During World War II, lifeboats and crates of clothes and food were washed up on the beaches from torpedoed ships in the Caribbean.

Fig. 50. Miskito woman cooking at her stove (*kubus*).

Fig. 51. Some of the implements used in cooking and food preparation. In the foreground there are gourds, tin cans, and an oil drum lid; in the corner, upside down, is a mahogany mortar.

drinking vessels. Most families have a few tin dishes and cups to eat from and perhaps a few forks and knives. One of the main implements used in the kitchen is a machete, usually one which was broken in the fields and reshaped.

All meats are eviscerated and vegetables peeled before cooking. Green bananas, plantains, cassava, breadfruit, *duswa,* dasheen, are all usually boiled—"lone water cooking." One of the diagnostic characteristics of Miskito cooking is the number of porridges (*wabul*) made from various vegetables and fruits, especially green bananas. After boiling, either bananas, plantains, breadfruit, dasheen, or green mangoes are beaten smooth with a small wooden pestle (*wabul dusa*) and coconut milk is added (Fig. 52). The Miskito say that *wabul* gives a quick feeling of fullness and strength. Wabul is often eaten in the early morning before going out to work in a swidden. There is no doubt about the fullness feeling as individual portions average one-half to three-fourths a quart.

Fig. 52. Making *wabul.* The porridge of bananas and coconut milk is beaten smooth with a pestle.

I suspect that the quick energy feeling results from the rapid liberation of sugars from already partially broken-down green fruits.

Food is cooked quickly without elaborate preparation. The excellence of foods in Tasbapauni comes as much from good tasting "raw materials" as it does from cooking methods. Stewed turtle fins, roasted paca, and manatee make for some of the best eating anywhere. Turtle is prepared in a number of ways, including boiling, roasting, and stewing. It is a multiflavored meat depending on the cut, thus greatly adding to taste variety. Mixed turtle meat is first boiled, and then is cooked with coconut oil.

Coastal Miskito use coconuts much more in cooking than do riverine Miskito. "Fish and breadkind" and "run-down" are two favorite foods receiving much of their flavor from coconut oil and coconut milk. Both are made with a base of different boiled ground provisions and breadkind (*duswa*, cassava, dasheen, breadfruit), which is then cooked in coconut milk; fish cooked in coconut milk is added for "fish and breadkind," and dried and salted white-lipped peccary or deer meat cooked in coconut oil and coconut milk is used to make "run down." Coconut milk and coconut oil are common ingredients in many Miskito foods, supplying a rich taste while helping to satisfy the Miskito's craving for fat. The meat of the coconut is almost never eaten. After being squeezed for its milk and cream it is fed to the chickens and pigs.

Another type of food for which the Miskito are known are sour foods, mainly partially fermented vegetable foodstuffs. Most of the sour foods are created by boiling vegetables (usually cassava, bananas, or pejibaye), beating them in a mortar, then allowing the mixture to stand for a few days. Soured food may be made into a drink, a gruel, or roasted. The final product can be kept for several days and serves to decrease inequities in food availability. *Bisbaya* is made by fermenting bananas wrapped in leaves and submerged in water; *bunya* is a soured vegetable paste from cassava or pejibaye which is made into a drink. Other soured foods, seldom made anymore, include *swanilaya* and *aungnilaya* made from fermented raw cassava and *biaralaya* from sugarcane juice.

Beverages used at the table are most often a bark or bush tea;[2] "Atlantico" coffee; *ibo* drink, a delicious nut-flavored beverage made from *ibo* nut paste; and water. Occasionaly "Chiripa," a flavored drink base bought at the store, may be drunk, but this is a special treat. Soft drinks and bottled beer are available and purchased at times when a little extra money is available. Homemade alcoholic beverages are no longer very

---

2. Most teas are made from locust bark, orange leaf, fever grass, soursop leaf, or ginger. This is only a partial list as the Miskito know and use many more roots, barks, leaves, and berries for teas, generally for medicinal purposes.

prevalent. In the old days the Miskito were famous for their "big drunks," the all day and night (and frequently the next day) drinking parties. Large quantities of *mishla* (a beer made from fermented cassava) were once consumed. Whole fields of cassava were dug up, the roots peeled, and placed in a *dori*, and pounded into a mush. In a short time a mildly intoxicating drink was obtained. Pressure from the churches and government officials has reduced *mishla* drinking to very rare instances. In some Creole and Black Carib settlements a distilled bush liquor called "kasusa" is made from sugarcane or granulated sugar. This drink and its distillation technology were introduced to the Miskito. Black Caribs in Orinoco, located 15 miles southwest of Tasbapauni across the lagoon, are locally famous for their "kasusa," some of which reaches Tasbapauni. However, social pressure against excessive drinking in the largely church-going Miskito communities have resulted in very little manufacture of alcoholic drinks.

Various reptile eggs are considered a delicacy, and are much sought after. A high protein food between February and June is provided by egg-bearing green turtles, occasional hawksbill nests uncovered on the beach and on the cays, iguana eggs from the beach and river sandbanks, and hicatee, caiman, and crocodile eggs from sandbanks and river edges. Eggs are usually boiled and spinkled with salt to protect against spoilage until eaten. All of the eggs are delicious, but perhaps the best are the ping-pong ball-size, light pink hicatee turtle eggs. On the other hand, chicken eggs are very rarely eaten; instead they are left for hatching to produce more chickens for possible sale in Bluefields.

A variety of breads and buns are made from bleached white flour purchased in the stores, which is mixed with baking powder, a little sugar, and some coconut oil.[3] Flour items are eaten in the morning when supplies of meat and ground provisions from the day before are very low or depleted. Bread is carried by turtlemen, hunters, and by agriculturalists on trips away from the village. The Miskito consider bread and other flour-based foods a necessity. Those Miskito who can afford flour may be favored with something to eat in the morning. Many Miskito, as we shall see, eat very little in the morning and almost nothing until they return from hunting or from their plantations.

The rather mild tasting plant food core of Miskito diets is frequently seasoned by using sugar, black pepper, *kumalaya* (pepper sauce made from *Capsicum* peppers), coconut oil and coconut milk, and various wild plants. A built-in variety is also provided by sundry wild meats obtained.

3. In Tasbapauni, the following foods were made with flour: buns, johnny cake, *tortilla* (bread dough fried in coconut oil), soda cake, dumplings, fritters, plantain tart, and "toe toe" (sweetly spiced biscuit). Coconut milk is used in all the breads and buns.

Cooking is considered a woman's work. Young girls help their mothers in the kitchens, frequently rotating cooking duties. Food preparation and cooking take about 3–4 hours per day. Most of the remaining housework involves making coconut oil and washing clothes. If a woman is to accompany her husband to the plantation she will postpone cooking until they return between noon and 2 or 3 P.M., depending upon the remoteness of the plantation.

FOOD STORAGE

Food storage and preservation methods are important for conciliating intermittent food procurement with constant food needs. As mentioned previously, the mainstays of Tasbapauni subsistence have built-in storage capabilities. Harvesting of plant foods occurs over a long period of time and root and tuber crops will keep if left in the ground. Protein supplies are mostly obtained from green turtle which can stay alive 2 weeks or more if kept under shade. The two major subsistence foods in Tasbapauni, turtle and cassava, are available for many months of the year and are capable of being "stored" for short periods; thus, they reduce the necessity for complicated and elaborate food storage and preservation.

It should be reemphasized that the food-sharing system in Miskito communities also has certain storage capabilities. Each food-sharing social link represents a latent "food cache," which can be activated according to needs and supplies of the involved parties.

Most of Miskito food storage and preservation efforts are for meat which spoils rapidly in the humid climate if not eaten quickly.[4] If the amount of meat to be saved is small it is cut in strips and smoked over the kitchen fire (Fig. 53). Larger quantities are often parboiled and salted or sun-dried, depending on the time of year.[5]

Inventories of preserved meat in houses are usually low or depleted. Meat distribution to kin and market sales often leave the hunter or turtleman with only a few pounds for his family. What is not eaten one day will be consumed the next. The nature of Miskito society and the present focus on cash sale of meat surplus, precludes large stockpiling of meat. It is only at the end of the dry season, in April and May,

4. Meat receives more care in preparation and preservation and more attention and circulation in reciprocal and market exchanges than any other type of food eaten by the Miskito. Much of the significance of meat and attitudes toward it among the people of Tasbapauni probably stem from the time when the Miskito were hunters and fishermen and only cursory agriculturalists.

5. Bennett (1962:47) describes similar meat preservation techniques used by the Bayano Cuna of Panama.

Fig. 53. Smoking white-lipped peccary meat. The animal's head is being roasted for the noon meal.

that some families will begin to sun dry and smoke meat and fish for the coming hunger months of June through August.

During April and May when meat is plentiful, some families, following traditional subsistence practices, start to prepare for the meatless rainy months ahead. Strips of turtle meat, turtle fins, and split, boned fish are hung from lines to dry in the sun. Because of pressure to distribute and sell meat at this time, no one ever seems to preserve enough. Older informants say the turtle crawls used to be built in the village where turtles were kept for months and fed on mangrove leaves until needed for food. Also, large quantities of meat were dried, but no longer. One fulfills kinship obligations, saves a little meat for the family, and sells the rest. In the end possibly 15 or 20 pounds a year will be preserved and stored. Even though meat will not be available in quantity during the heavy rain months, by selling and sharing people maintain an incoming trickle of money and avert social frictions over large meat stores. If a number of men sell out their meat catches rather than preserve portions for the rainy season, increased pressure will be placed on those who save meat until their supplies are also depleted.

Many traditional plant foods which were previously preserved for storage are rarely made today due to the availability of purchased foods.

For example, *konkanti,* a flour made from pounded dried slices of bananas, is seldom used because bleached white flour can be bought in the village. Market purchases are supplanting food storage as the means to guarantee food availability. The problem of course, is how to guarantee the availability of money for these purchases, and once made, is the purchased food of equivalent nutritional quality?

FOOD CONSUMPTION AND NUTRITION

The amount and frequency of food consumption reflect adjustments to food procurement and distribution patterns as well as a cultural conditioning of appetites and digestive habits. Nutritional value is important for the health, vigor, and energy potential of the food consumers. Both food consumption habits and food nutrition are significant indicators of the success of a group's ecological adaptation to a particular locale.

Collecting data on food consumption and nutrition was the most difficult part of our fieldwork. Food is a personal part of a person's or a family's life. For many groups and peoples it is the center of their life. We found that three families were the maximum sample we could handle for dietary analysis and still keep on with other research objectives. The families were not selected at random. As is the case in many field studies, sample families or informants are those who are at least tolerant to the fieldworker's strange probes and lifestyle. One does not want to disrupt eating habits or the types of food normally consumed. For this reason it is imperative that the sample families feel somewhat comfortable with the observers and realize that they did not come to eat their food. Also, as snacking on a variety of foods is commonplace, one must stay with the informants as much as possible or make allowances for estimating snacks.

Dietary measurement was carried out during a 14-day period in the dry season and another 14-day period in the wet season. Additional measurements were made at intermittent times for 1 or 2 days throughout the fieldwork. For the 28 days total of intensive dietary analysis, specific data depicting food quantity and quality was obtained. More general dietary information from scattered 24 to 48 hour diet descriptions help to portray the over-all annual form and pattern of Tasbapauni Miskito diets.

All foods brought into the three families' houses were weighed during the diet study periods. Foods were weighed fresh and cooked. If individual portions could not be weighed, estimates were made for the amount each person received. Amounts of meat and vegetable waste were obtained by subtracting foods ready for cooking from their weights when brought into the house. Measurements and estimates for foods eaten away from the houses were added to these figures. Nevertheless, because of snacking in swiddens, in others' houses, and at odd times of the day, it was impossible to accurately record the weights and kinds of all food eaten. Probably 5% or so of the food eaten was not recorded. For every food consumed during the day, nutritional values were recorded based on available tables,[6] or, if a particular food has not been evaluated, approximations were assigned. Nutritional values were usually derived from fresh, uncooked foods.

Dietary and nutritional measurements could of course be carried out much more accurately by trained researchers with laboratory facilities for on-the-spot analyses. There appears to be little interest among nutritionists to work with Indian peoples in such remote areas.[7] So one does what one can.[8]

Consumption

Eating food is more than simply an ingestive process. It is an act involving a whole complex of cultural, social, and physiological arrangements. Food consumption patterns express a group's interrelationships with their food procurement, distribution, and preparation systems. Much of one's appetite and digestive habits are conditioned by upbringing within a particular cultural context. This is also important in determining when and in what quantities food will be eaten.

Meal times in Tasbapauni are very irregular. The largest meal of the day is eaten around noon. If available, large amounts of meat and ground provisions are consumed. In the morning a small meal of bark tea and bread might be eaten at 6 or 7 A.M. Men often skip breakfast if they leave the village in the very early morning to go to their plantations, hunting, or fishing. Sometimes they carry a few bread rolls and some cold coffee, or stop to cut a young water coconut.[9] Evening meals are small as were breakfasts and usually consist of bread and coffee or leftovers from the midday meal.

6. The most helpful nutritional tables were by Wu Leung (1961) and Bradley (1956) which list many native plants and animals.

7. I was told by more than one nutritionist in Central America that dietary problems among Ladino peasants and city dwellers are far greater than with most Indian people; hence there is little dietary research on Indians.

8. My field procedures for diet measurement closely paralleled Rappaport's (1968:278–284) efforts among the Tsembaga Maring of New Guinea, although I did not become aware of his study until I returned from Nicaragua.

9. A water coconut is a small immature coconut with very sweet water and thin jelly meat, not like mature coconuts which have almost acidic water and thick coconut meat.

Daily food intake is highly variable in timing and amounts. Using average figures for calories consumed over a period usually obscures this fact. Though Miskito food consumption is concentrated on one large meal per day, even this is extremely irregular as to time of day because of varying labor demands in food procurement and changing weather conditions. When the weather breaks one must be ready to go out to seek food or to care for a plantation. Good weather is to be taken advantage of; one may not even come home for the midday meal. Hunger is an accepted condition. It can be endured for long periods.[10] Yet when there is much food, people eat large amounts and eat often. It is the Miskito's adaptability or tolerance to fluctuating food supplies which most strikes the outside observer. High tolerance of hunger and a large capacity for eating are regulatory physiological and cultural mechanisms which act as a buffer counteracting inconsistencies in food procurement and availability.

Men usually eat first, women and children afterward. Some families have a table to eat on, while others do the best they can in hammocks and on stools and benches. Little conversation takes place; eating occupies everyone's attention (Fig. 54).

Fig. 54. Eating the noon meal—cassava and turtle meat.

10. Turtlemen and hunters say that smoking cigarettes helps dull hunger and allows them to travel light and fast without having to make stops to "look food."

Nutrition

There are many difficulties in attempting to assess the nutritional composition of an Amerind group's diet. First of all very little work has been done on aboriginal diets in Latin America and a good deal of what is known verges on speculation. Bennett (1967) pointed this out in reviewing ecological work in Middle America:

> Our knowledge of dietary requirements is too often based upon assumptions extrapolated from studies of nonaboriginal populations. Research has too often proceeded according to dogma established by an "authority" rather than by first-hand observation and, when possible, experimentation [p. 18].

It is one thing to measure the nutritional makeup of a group's diet and quite another to gauge it according to some minimal dietary requirements stipulated by midlatitude Western culture criteria. Newman (1961) concluded that there are many groups who consume but a fraction of the supposed minimum requirements. He suggested that some human populations have developed a tolerance to subminimal nutrient intake which would be harmful to other populations.[11] Hegsted (1954) argued that "one may reach the conclusion upon theoretical grounds that the amount of a nutrient required to maintain balance in a normal adult is that amount which his traditional diet supplies [p. 105]." For example, even though the traditional Eskimo diet of mainly meat, fish, and fat contains no fruit, vegetables, milk, or eggs, the Eskimos appear to be getting along quite well.[12] Furthermore, factors such as intestinal parasites and constraints on the metabolic availability of nutrients[13] may seriously limit any interpretations of the adequacy of a particular diet.

As Whiting (1958) noted: "The most commonly used and perhaps the most important single measure of food intake is the adequacy of caloric level [p. 73]."[14] Of the diets sampled in Tasbapauni the average daily caloric intake for an adult male ranged from 2000 in July when all foods are scarce, to 2400 in May when crops are scarce and turtle abundant, to 2800 in September and October when the first cassava

---

11. For more on nutritional adaptation see Lasker (1969:1483–1484), Alland (1970:68–85), and Newman (1961:627–629).

12. See Whiting (1958) and Nelson (1969:158; 180–181).

13. Nicholls (1961:336), in his book on tropical nutrition and dietetics, listed several factors in food preparation and composition which may act to limit the full use of ingested foods and their nutrients.

14. Nutritional recommendations suggested for Central American populations by the Instituto de Nutrición de Centro América y Panamá (1966:75–76) for a male of 120 pounds are 2700 calories and 65 grams of protein daily. The amount of fat or fatty acids desirable in a diet has not yet been established. The average U.S. fat intake is 40% of the total calories consumed (Goldsmith 1966:164).

crop is ready and green turtles return from nesting. The daily average for a 1-year period was approximately 2500 calories. As stated previously, daily caloric intake was extremely variable; some days a man would consume barely 1000 calories, or even only about 700.[15] When food became plentiful, daily consumption levels rose over 4000 or 5000 calories. During the dry season daily calorie intakes were higher and less variable than in the wet season.

The Miskito are not waterside people by accident. Their preoccupation with meat-getting activities and the significance of meat in Miskito society and diets is greatly facilitated by their river and coastal location. The availability of protein is especially important for the coastal Miskito because their crop staples are mostly high starch roots, tubers, bananas and plantains. During the dry months about 65–70 grams of protein (largely of animal origin) were consumed daily by adult men; in July only 30 grams were ingested. Protein intake is more variable than calorie consumption, however, averaging 50–60 grams daily for the year.

The Miskito say they feel strong and full right after eating meat.[16] Consumption of large amounts of protein-rich meat, while adding little in the way of calories, staves off hunger for a long period, and apparently gives an immediate sense of well-being and strength.[17]

It has often been stated that high protein diets in the tropics were disadvantageous because they cause high metabolism and body heat;[18] however, this argument is blunted by the fact that there are many tropical peoples with a high protein diet. Experiments made in 1957 on the metabolic effects of protein showed that in terms of heat disposal a high protein diet itself was little different from a low protein diet of equal calories and that there was no reason why protein-rich foods should be avoided in hot, humid weather (Swift 1959:52–55).

The degree of protein utilization by the body is determined in large part by the kinds and proportions of amino acids. Proteins derived from meat contain most of the essential amino acids, permitting their almost complete use. A diet containing a wide variety of foods offers a greater

---

15. The frequent Miskito medical complaint of faintness and dizziness may be associated with periods of subminimal food intake. In the United States, sensations of faintness were reported by individuals being tested under semistarvation (1000 calories per day) laboratory conditions. The subjects also registered complaints of light-headedness, dizziness, and weakness" (Borzek 1966:54–55).

16. After meat ingestion an increase in basal metabolic rate occurs which is greater than the intrinsic caloric value contained in the meat. This stimulus is called the "specific dynamic effect" (Nicholls 1961:23).

17. A similar description of strength-giving qualities of meat is given by Richards (1961:58–59) for the Bemba of Africa. According to Nelson (1969:179) the Wainwright Eskimo believe that meat eaten with fat or oil will heat the body and help to maintain body warmth.

18. See Nicholls (1961:23–24) and Vermeer (1964).

opportunity to provide additional or lacking amino acids for protein utilization. The variety of traditional Miskito foodstuffs encourages the availability of essential amino acids, thus increasing the percentage of ingested protein which can be absorbed by the body.[19] Large amounts of essential amino acids are supplied in animal organs, complementing and supplementing those found in muscle tissue. The Miskito's habit of insisting on eating "mixed meat" definitely improves the utilization of protein intake. Dietary simplification, or the loss of variety of food items in the diet, may act to reduce the availability of some essential amino acids.

Through their intermittent but high consumption of meat, the Miskito are able to achieve quick recovery from periods of low calorie intake. In addition, the generally superior health and physique of the coastal Miskito compared with the riverine Miskito may be because of the generally higher protein intake in coastal villages. Alland (1970) states that "there appears to be some relationship between protein intake and resistance to infectious disease. Antibody production is related to protein intake because the basic building blocks of antibody molecules are the essential amino acids [p. 76]."

Second only to the craving of meat is the Miskito's desire for fat. Meat is good say the Miskito, but fat meat is better. An additional major source of fat is coconut oil and coconut milk, important ingredients in many Miskito dishes and cooking techniques. The daily intake of fats in the sample diets of adult men in Tasbapauni was 70 grams in May, 40 in July, and averaged 60 grams for the year. Rainy season hunger months for the Miskito do not exist due to lack of food, which averages 2000 calories per day, but because of the reduction in the amounts of protein and fat in the diet. The Miskito complain of becoming tired quickly when there is little meat and fat to eat.

Fats are the major energy producing item in Miskito diets. They are a much more concentrated form of energy (nine calories per gram) than are carbohydrates and proteins (both four calories per gram). Approximately 18% of the total calories consumed for a sample adult diet in Tasbapauni come from fats. Fat and protein provide long lasting energy foods. Rappaport (1968:136–137) reports that shortly before fighting, Tsembaga Maring warriors in New Guinea usually ate salted pig fat which provided them with a burst of energy in about 2 hours. The Miskito often try to eat some meat during periods of stress (a death, funeral) or before doing a particular difficult task, or before a long journey.

19. See also Rappaport's (1968:74–76) discussion of Tsembaga Maring dietary patterns and available amino acids.

The importance of fatty meat in the diet of the Eskimo and of the Northern Athapaskan Indians has been discussed by Stefansson (1960:36). He reported that fat deficiency often leads to headaches and can, after 1 to 2 months, lead even to death. Similarly, Rohrl (1970:97–101) suggests that the cause of the windigo psychosis among the Chippewa, Cree, and other Northern Algonkian peoples may be the result of a deficiency of adequate amounts of fat and associated fatty acids and vitamins in the diet. For the Miskito the high fat intake probably serves a number of functions; it quickly dissipates any feeling of hunger, makes a meat more palatable in the eyes of a Miskito, helps overcome fatigue from early morning plantation work or other food procurement activity, and provides a source of quick energy for women whose labor demands are more constant than the short and intense labor efforts of the men. In fact, at the turtle butcherings it is the women who yell the loudest for a piece of fatty meat.

The relatively high intake of calories, protein, and fat and the generally adequate supply of vitamins and minerals from the varied food sources utilized by the Miskito of Tasbapauni observed and measured during 1969, contrast with the situation in many other Miskito communities. Even though formation of food-getting systems are influenced heavily by specific cultural behavior and technology, local differences in the types and amounts of available resources can elicit concomitant differences in dietary patterns, food cravings, and cultural and possibly physiological adjustments. Thus, on the Río Coco, where diets are much more deficient in animal protein than on the coast, beans assume a major protein-yielding role in dietary patterns. Riverine Miskito are typically much smaller in body size and stature than coastal Miskito, but this may be as much the result of the latter's high admixture of foreign genes as due to their superior diet. Calorie, protein, and other nutrient and vitamin deficient medical cases are much more prevalent for riverine Miskito than for coastal Miskito.[20] One rarely sees symptoms of kwashiorkor in coastal villages, but along the Río Coco it is much more common. Although superior to diets based on interfluve resources, riverine diets do not match the quality and quantity of most diets of the coastal Miskito. Tasbapauni's favorable location between the forest and the sea allows multidirected exploitation efforts, thus favoring food diversity, spreading out subsistence risk, and increasing the chances for dietary balance.

The measured percentages of calories obtained by the three families sampled in 1969 from different means of food procurement roughly are: agriculture 74%, purchased store foods 18%, hunting and fishing

20. Personal communication with Dr. Edwin Wallace, Puerto Cabezas, Nicaragua and Dr. Peter Haupert, Bilwaskarma, Nicaragua, 1969.

7%, and gathering 1%. The approximate daily consumption of calories and protein for men, women, and children is presented in Table 31. The totals were derived by extrapolating average consumption figures for the sample families.

TABLE 31

Estimated Daily Consumption of Calories and Protein for Individuals, and for Tasbapauni, October 1968 through September 1969

	Individual				Tasbapauni		
	Calories	Animal protein grams	Vegetable protein grams	Number of individuals	Calories	Animal protein grams	Vegetable protein grams
Adult males	2500	30	23.2	178	445,000	5340	4130
Adult females	2200	26.4	20.9	165	363,000	4356	3449
Adolescent males	2000	24.0	18.6	112	224,000	2688	2083
Adolescent females	1900	22.8	17.6	143	271,700	3260	2517
Children 6–10	1250	15.0	11.6	183	228,750	2745	2123
Children 2–5	1000	10.0	11.3	150	150,000	1500	1695
Children 1 year	800	?	7.0	35	28,000	?	245
Children less than 1 year	?	—	—	31	?	?	?
				997	1,710,450	19,869	16,242

Based on these data an average family of seven, including an adult man and woman, adolescent girl and boy, one child 6 to 10 years old, and two children 2 to 5 years old, would consume approximately 12,000 calories per day, and about 4,380,000 per year; 138 grams of animal protein or about 1½ pounds of meat per day, and 548 pounds per year. Average daily consumption rates for the village of 997 people work out to 19,869 grams of animal protein (roughly 220 pounds of meat), not all of which is equally distributed. Compared with other Miskito villages, Tasbapauni animal protein intake is high and calorie levels are about the same.

These data do not point out at least three important maladaptive trends operating toward decreasing the quality of Tasbapauni diets: (1) reduction in the frequency and amplitude of reciprocity in food distribution within the village, especially for meat, so that marginal producers are becoming marginal receivers; (2) simplification of the subsistence base due to market participation, causing irregular and marginal food availability; and (3) increasing dependency on generally low nutritional quality market foods in lieu of higher quality traditional foods.

The limited dietary analyses done in 1971 indicated that since the intensification of market sale of local subsistence resources, less meat was being eaten in the village, there were greater discrepancies in food consumption between individuals, and more reliance was being placed on purchased processed foods than was the case in our earlier work. At this stage these are generalizations needing detailed measurement for documentation. Nevertheless, if these maladaptive trends continue there will be serious and widespread nutritional problems among large numbers of the population.

In order to place the 1969 Tasbapauni dietary patterns in perspective it might be well to compare them with those of another Central American Amerind group, the Bayano Cuna of Panama (Table 32). The data for the Miskito example were obtained from observations of three adult

TABLE 32

Nutritional Components of Average Tasbapauni Miskito and Bayano Cuna (Panama) Diet, Bayano Cuna Statistics from Bennett (1962:46)

	Approximate daily amount grams	Percentage of total weight	Protein grams	Fats grams	Carbohydrate grams	Calories
Tasbapauni Miskito						
Cassava	600	40.0	6.0	2.4	196.8	792
Bananas/plantains	200	13.3	2.3	0.3	61.2	231
Duswa	200	13.3	3.4	0.6	61.8	264
Meat ("mixed" green turtle)	150	10.0	30.0	8.0	—	166
Fruit (in season— mango, pineapple)	100	6.7	0.4	0.2	14.0	55
Dasheen	100	6.7	1.6	0.2	22.4	92
Flour bread	100	6.7	9.5	1.0	65.0	450
Coconut oil	50	3.3	—	50.0	—	450
	1500	100.0	53.2	62.7	421.2	2500
Bayano Cuna, Panama						
Bananas	1800	89.0	21.6	3.6	414.0	1800
Maize	113	5.6	10.0	4.0	81.3	401
Meat	57	2.8	9.0	[a]	—	70
Fish	51	2.5	9.1	[a]	—	54
	2021	99.9	49.7		495.3	2325

[a]Not measured.

men during two 14-day periods and checked against food consumption at other times of the year.[21] It should be stressed that these figures represent consumption study periods. A man might eat 600 grams of green bananas, boiled or in *wabul,* one day and not eat any more bananas for several days. Some major differences are readily apparent in the Bayano Cuna and the Miskito sample diets. Although total calorie intakes are relatively similar, amounts of animal protein in the Miskito diet are almost double that for the Bayano Cuna. Moreover, almost all of the Bayano Cuna calorie intake came from two sources, bananas and maize, while Miskito calorie sources are more diversified. In general, the Miskito diet appears to be nutritionally superior to the Bayano Cuna diet. Some interesting results might be obtained by comparing dietary patterns of the Bayano Cuna, a riverine people in Panama, and the San Blas Cuna, a coastal group, with their Miskito counterparts in Nicaragua.

21. Miskito beverages are not included in the table because most of them (water, various teas, coffee) contain few calories other than those contained in any sugar used. *Wabul* consumption has been included with "bananas and plantain."

CHAPTER X

SUBSISTENCE STRATEGY, REGULATION, AND CAPACITY

Miskito subsistence is a system made up of functionally related components and interrelated ideas and activities. The goal of the subsistence system in Tasbapauni is to provide adequate amounts of culturally acceptable foodstuffs throughout the year to the villagers. The subsistence system functions to adjust the irregularities and inconsistencies of the resource base to the regularity of nutritional demands. Because of differing crop seasons and availabilities of various biota, the Miskito developed a subsistence strategy to increase the likelihood of food procurement by lowering subsistence risk and allowing a wide mobility of food-getting pursuits by different means, in different places, and at different times of the year. The diverse and wide-ranging subsistence strategy diluted exploitation pressure over time and area so that severe ecological disruptions were minimized while assuredness of subsistence was maximized.

The system could be internally regulated to maintain a subsistence goal range within ecologically tolerable levels. The subsistence system was adapted to environmental flux in the availability of biotic resources and changes in weather and sea conditions. Maladaptive trends and factors are presently disrupting local control of subsistence resources and strategies. For the most part these disruptive factors are being generated from external systems.

223

First let us take an overview of the system and how it functions before moving on to explain variations, strategies, regulation, adaptation, the impact of the system on the environment, and the production and capacity of the system.

In order to look analytically at a system a certain amount of abstraction and closure is necessary to permit measurement and description. As Harvey (1970) points out: "In reality any system is infinitely complex and we can only analyse some system *after* we have abstracted from the real system [p. 448]."[1]

For the purposes of explanation and examination, the Miskito's system was abstracted on the basis of identifying: (1) a set of elements identified with some variable attribute, (2) a complex of relationships between the attributes of objects, (3) a set of relationships between those attributes of objects, the population, and the ecosystem, (4) and an interpretation of the cognitized system.[2]

The Miskito's subsistence system is traditionally composed of a set of different resource procurement activities organized for sustenance, with socially regulated production and distribution, adapted to environmental and ecological factors. The structure of the system is primarily made up of elements, links or relationships, and a goal range. At the level of abstraction shown in Fig. 55, elements are represented by such things as the different ecosystems and the cultural and ecological constraints involved in the system. At a finer stage of abstraction each of these elements can be broken down into subsystems of their own. Resources, decisions, and behavior, on the one hand, and environmental processes on the other, act as links between each of the elements connecting the system into a functioning entity. It follows that if the system is composed of elements interconnected by functional relationships, a change in one of the elements or relationships would cause a change in the others.

Within each of the broadly defined ecosystems exists a range of resources which the Miskito define as food. A series of compound relationships operate between resource availability and accessibility and labor demands in other subsistence activities. From this complex, traditional

---

1. Similarly, Hagen (1961) noted:

> For use in analysis, a system must be "closed." A system which is interacting with its environment is an "open" system: all systems of "real life" are therefore open systems. For analysis, however, it is necessary in the intellectual construct to assume that contact with the environment is cut off so that the operation of the system is affected only by given conditions previously established by the environment and not changing at the time of analysis, plus the relationships among the elements of the system [p. 145].

2. Based, in part, on Harvey's (1970:451) interpretation of a system.

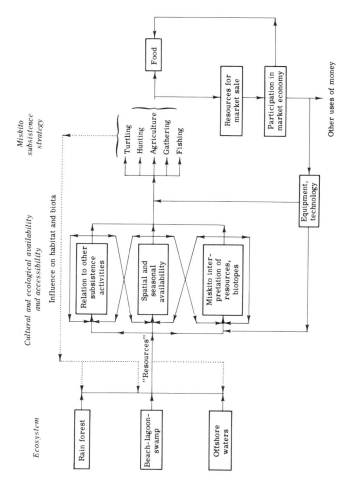

Fig. 55. Generalized diagram of food getting and interrelationships.

225

Miskito subsistence strategy selects various combinations of food-getting methods based on their ethnoclassification of food, with an attempt to lower subsistence risk and acquire a relatively assured and balanced diet.

The net result of subsistence efforts is food and, secondarily, resource items which can be converted into money, food, or other goods. Some of the money is fed back into the system by way of equipment and directly affects Miskito subsistence strategy and resource evaluation. Feedback from subsistence efforts themselves work in turn as selective forces on components in the environment, especially on specific fauna and flora.

STRATEGY AND STRUCTURE

The Miskito have a number of alternative means of procuring food. They can cultivate a swidden, gather wild foods from the forest and the sea, go hunting and fishing, or buy food with money earned from the sale of agricultural, forest, or sea products. These means of food getting are not mutually exclusive but are combined into one overall food-getting system for each family, kinship group, and for the village of Tasbapauni as a whole (Fig. 56). The interrelationships, combinations, and relative importance of different strategies of production for use and production for exchange have differential returns and risks.

In Fig. 57 a portion of the subsistence system is diagrammed in somewhat greater detail. In this diagram the flow of money, food, and labor is shown for one typical nuclear household from food procurement through food preparation and consumption. Even though the diagram was simplified greatly, it still shows a complex network of elements and relationships, each requiring a specific action or decision to activate it. Before analyzing different strategies, it is necessary to examine the productivity of each food-getting activity in order to arrive at some bases for comparison.

In the preceding pages an effort was made to describe each means of food procurement in terms of time and distance inputs and yields outputs. Much of the data were very crude and incomplete. However, productivity ratios (hour of labor to calories and pounds produced) were arrived at.

The use of calories as the common denominator for comparison of different crop yields, or for different meat yields from hunting and fishing, has many obvious limitations. First, accurate calorie values per measured unit do not exist for all of the foods eaten by the Miskito;

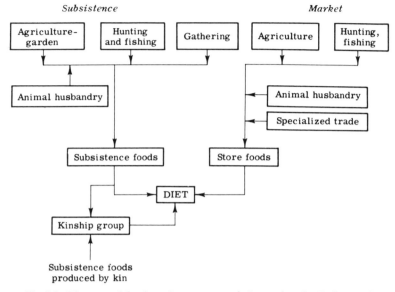

Fig. 56. Diagram of food-getting means and alternatives in Tasbapauni.

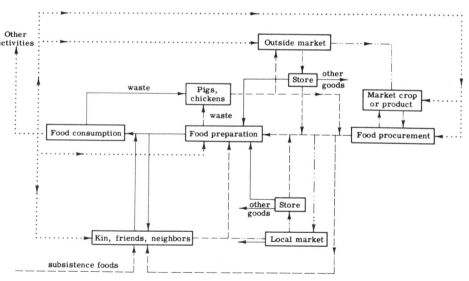

Fig. 57. Generalized diagram of the circulation of foodstuffs, money, and energy-labor for one nuclear household of Tasbapauni. (—–): Unprocessed food; (———): Processed food; (–·–·): Money; Energy-labor (· · ·); Market item (–··—··—··).

therefore approximations had to be made. Second, the idea of using calories assumed that there is a standard percentage of each crop grown for all of the plantations. If a person, for example, grew nothing but high calorie, high yield cassava, instead of only approximately 50% of the total land cultivated, then his calorie productivity would be very different. But no one plants all cassava, nor does anyone buy all flour at the store. Another problem is that there are obviously different degrees of work effort involved in each food-getting task. An hour of walking through a swamp is harder than sailing a *dori* after turtle. The best way to compare systems and work loads would be to have data on energy expended in calories per task, per individual, rather than using man-hours as I have had to do.

And, finally, the concept of reducing everything to calories ignores the respective food value of different foods. In other words, Miskito subsistence efforts in turtle fishing may have a low calorie productivity to time invested compared with agriculture, but the protein return is much greater. Nevertheless, the use of calories does provide a common denominator for comparison of the different systems when these limitations are taken into account.

The amounts of calorie return in sampled subsistence and local market activities are presented in Table 33. Either by using total calories produced per hour of labor or by assigning 175 calories per hour[3] as an average energy expenditure rate, it can readily be seen that agriculture has a calorie productivity many times that of other types of food procurement. Miskito subsistence agriculture not only has a high calorie ratio (calorie return:calories expended in labor) of 30.4:1, it also has a fairly dependable calorie return. Comparing other subsistence activities shows that calorie ratios for hunting (7.1:1) are higher than turtle fishing (5.5:1) or gathering *ahi* (2.1:1), due to the hunting focus on high calorie yield white-lipped peccary. Food getting by way of participation in the market economy gives the second highest calorie ratio (11.5:1), substantially augmenting subsistence efforts and reducing subsistence risk by decreasing the amount of food a family must procure directly; however, procurement of cash market resources may disrupt the timing of subsistence activities and thus lower their efficiency. The amount and degree of

---

3. The rate of energy expenditure is dependent primarily on body size and the nature of the activity. Extrapolating from Hipsley and Kirk (1965) and Swift (1959:49), I have chosen an average energy expenditure figure of 175 calories per hour for a man of 150 pounds (an average Tasbapauni male) working at medium labor. Comparative calorie per hour rates are: walking at 3 miles per hour, 140 calories; walking at 4 miles per hour, 238, and running, 490. Rappaport (1968:258–259) estimated that a 120-pound man in the New Guinea Highlands clearing underbrush (medium-heavy work) expended 195.6 calories per hour over basal metabolism.

TABLE 33

Approximate Calorie Return per hour Input of Work and for Calorie Input for Different Means of Food Acquisition, Tasbapauni, 1968 through 1969

Means of food acquisition	Calorie yield per hour	Ratio of calorie yield to labor input at 175 calories per hour	Animal protein grams per man-hour input
Subsistence agriculture	5335:1	30.4:1	—
Hunting	1251:1	7.1:1	136
Turtle fishing	956:1	5.5:1	173
Market economy[a]	2012:1	11.5:1	—
Gathering (*ahi* only)	372.1[b]	2.1:1	62

[a]Based on traditional buying patterns of store foods and their caloric value.
[b]Higher for many other gathered wild foods.

dependability that market-based foods can give to Miskito diets are tied closely to foreign market demands and prices for forest and sea resources exploited by the Miskito. Market demands and resource availability, however, can fluctuate greatly. Indeed, jaguars and ocelots are becoming vary rare and green turtle populations are declining. Therefore, the amount market exchange transactions can contribute to Miskito diets is sensitively adjusted to factors which the Miskito have little control over.

In addition to the importance of turtles in the village market economy and the already mentioned low subsistence risk of turtling compared with hunting, another reason for the dominance of turtling is the high protein return. As has been shown, hunting is generally more productive in calorie returns than turtling but agricultural calorie productivity is so great that their margin of difference is negligible. On the other hand, the grams of animal protein return per man-hour for turtling (173) is 20% higher than for hunting (136) and 278% greater than collecting *ahi* shellfish. Added to this the fact that protein return is also almost 20% more dependable in turtling than in hunting, it is evident that Miskito meat-getting strategy has considerably reduced subsistence risk and increased protein productivity by concentrating on turtling.

Investigators elsewhere have used calorie ratios in their analyses of subsistence. Rappaport (1968:52) found that the Tsembaga Maring of New Guinea had a calorie ratio of 15.9:1 for sugar–sweet potato gardens and 16.5:1 for taro–yam gardens. Marvin Harris (n.d.) estimates calorie ratios for the Dyak (Borneo) rice swiddens at 10:1 and Tepoztlán (Mexico) maize swiddens at 13:1 on poor land and 29:1 on good land. Leeds

(1961:23) reported that the Yaruro (Venezuela) had ratios only of 8:1 or 9:1, principally from cassava swiddens and from hunting. These estimates suggest that Miskito calorie ratios (30.4:1) for agriculture are relatively high, supporting the observation of high yields per hour of labor.

It is also necessary to look at how time is used or allocated between different activities in order to more fully understand the subsistence system. Using the figure of 8 hours per day as the average work day,[4] 2920 hours per year theoretically should be available for work. Based on day-to-day observations of three families it is estimated than an average of 117 days (936 work hours) per year are unavailable because of sickness, bad weather, and other factors (Table 34). This leaves only 1984 hours (248 days) for work. Adding an additional 1984 hours for a wife, and 1000 for two working-age children the total potential work hours available for an average family of seven is approximately 5000 per year.

Assuming that the foregoing productivity ratios are reliable, some 6,200,000 calories, or 2,700,000 more than the 4,500,000 annual calorie intake for a family of seven, will be obtained from an average input of some 2000 hours in agriculture, turtling, and market selling.[5] Other tasks about the home (food preparation, house repair) and village (*dori* repair, gathering firewood) occupy a significant part of the remaining available work hours. Many families may work more and produce more than this example, and others less.

Time expenditures in different activities are not proportional to return, amount of food distributed through reciprocity, nor contribution

TABLE 34

Loss of Potential Work Days of a Year. Estimate Compiled from Observations of Three Families, October 1968 through September 1969, Tasbapauni

	Days
Sundays—not supposed to work	52
Deaths in the village—stay home	10
Bad weather—cannot go out	25
Sick—cannot go out	20
New Year's, Easter, Christmas weeks—stay home	10
Total	117

4. This is 3 hours longer than the Miskito's stated 5-hour work morning, but the additional hours should compensate for the greater amount of time spent per day on turtling, hunting, and traveling.

5. The calorie total is made up of approximately 4,500,000 from subsistence agriculture, 500,000 from hunting and fishing, and 1,400,000 from local market foods. See pp. 219–220 for explanation of 4.5 million calories consumed by a family of seven.

to the diet. The following are rough estimates of average time expenditures for different activities for a Miskito male during daylight hours: (1) 35% spent on production for use, (2) 28% on production for exchange, and (3) 37% on leisure, eating, and small-scale repair of equipment. Subsistence activities do not take much time; they are relatively high yielding for low labor inputs. During a year period almost as much time is spent on cash market efforts as on subsistence agriculture but they provide only 18% of the diet (calories) while subsistence agriculture yields 74% (Table 35). Subsistence activities also have a high distribution frequency through reciprocity (an average of 25–30% of the total produced) while items secured by cash have a very low distribution frequency outside of the nuclear family (5%). By reducing time inputs and by limiting production for reciprocity in subsistence, both labor and resources can be diverted into money-yielding channels. Livelihood strategy in Tasbapauni is tending toward an attempt to gain a maximum yield with a minimum time input in subsistence, and utilizing surplus time for market efforts. A time-bind is frequently created when individuals try to playoff subsistence and market opportunities and end up with rescheduled labor inputs that are out of phase with environmental and ecological factors. Thus, "efficiency" (yield for time input) is sometimes reduced while intensity of exploitation (time input on a species or site) goes up. Even though possibly less efficient, more intense exploitation can be extremely disruptive ecologically if it greatly exceeds the ecosystem's capacity for reestablishment and ability to stabilize.

In interacting with their ecosystem, the Tasbapauni Miskito are intensifying exploitation of market resources while simplifying the subsistence system. Thus their over-all adaptation in the system is becoming less diverse, less stable at the same time livelihood strategy is becoming more complex. The loss of diversity of subsistence channels effectively reduces the efficiency of Miskito adaptation. Margalef (1969), in writing about diversity and stability, observed:

> If we consider the interrelations between the elements of an ecosystem as communication channels, we can state that such channels function on the average more effectively, with a lower noise level, if they are multiple and diverse, linking elements not subjected to great changes. Then, loss of energy is lower, and the energy necessary for preventing decay of the whole ecosystem amounts relatively to less [p. 378].

REGULATION AND ADAPTATION

The Miskito subsistence system can be considered a complex system incorporating various mechanisms and characteristics for regulation and adaptation. System regulation is affected through processes by which

TABLE 35

Subsistence and Market Food Production and Distribution, Tasbapauni

	Distribution of foodstuffs; relative percentage for one family					Adult male, daily average			
	Nuclear family use	Distribution to kin and others	Sell	Calorie ratio[a]	Protein/man hr	Number of calories	Percentage of calories	Vegetable protein grams	Animal protein grams
Turtling	20%	30%	50%	5.5:1	173	120.0	5	—	21.6(72%)
Hunting	20%	30%	50%	7.1:1	136	27.0	1	—	4.8(16%)
Agriculture	70%	25%	5%	30.4:1	—	1880.0	74	13.7(59%)	—
Gathering	80%	15%	5%	2.1:1	62	20.0[b]	1	1.0(4%)	.3(1%)
Store purchases with money from market economy	90%	5%	5%	11.5:1	—	453.0	18	9.5(37%)	.6(2%)
Fishing	20%	30%	50%	?	?	20.0	1	—	2.7(9%)
						2520.0		24.2	30.0

[a]Calorie return:calories expended.
[b]Estimate.

systems maintain their structure, while adaptation is brought about by processes through which the structure of the system changes in response to environmental pressures (Rappaport 1968:241). Regulatory mechanisms are chiefly enacted through cultural and social channels in order to achieve a "preferred state" or preferred level of food production and distribution.

Both regulatory and adaptive system adjustments are influenced by system feedback, or the results of the system interacting on itself. In addition, the system can be influenced and changed by conditions from outside its normal parameters. In discussing regulatory means it should be noted that there are various degrees of sensitivity and ability in subsistence regulation. None of the Miskito's regulation mechanisms are very sensitive, capable of making rapid adjustments. Some of the system is self-regulating, with various components reacting to changing environmental conditions at different times of the year and in different places without a stimulus from the Miskito.

Traditionally, the Tasbapauni Miskito's subsistence system was regulated by the scheduling of labor inputs on diverse species and sites occurring over a large land and water area, by social controls of the intensity of production for use and distribution of materials, and by cultural perception and selection of resources. The system was adapted to the seasonality of resource availability and optimal environmental conditions for efficient exploitation, and to the highly diverse ecosystems with which the Miskito interacted. Production for use based on long fallow polycultural swidden agriculture and hunting and fishing of diverse land and water fauna, minimized ecological disruption while supplying relatively assured and balanced subsistence.

Present-day subsistence of the Tasbapauni Miskito still retains many of the traditional patterns and adaptations. Miskito agriculture is based primarily on crops which can be left in the fields for long periods with little attention, thereby allowing for pursuit of other subsistence activities, especially hunting and fishing. The occurrence of game animals in agricultural fields and of aquatic food resources adjacent to waterside plantations, permits the Miskito to concentrate diverse food-getting efforts on localized areas without having to increase time spent in travel. During the February through May dry season, labor inputs are high and subsistence activities are split between land and water. Turtle fishing is easy and quick permitting frequent visits to agricultural sites for clearing and planting. During the wet season subsistence efforts are turned toward land, distances traveled are reduced, and hunting becomes important.

Practically all subsistence resources and food-getting procedures have been selected to work in unison and to provide ancillary food alternatives if ecological conditions on land or water become unfavorable. Miskito subsistence is largely adjusted to cassava-based agriculture and to turtle-based meat getting. Miskito cultural attitudes toward the available food supply are partly adapted to these two relatively common foods. Both of these resources contain built-in buffers helping to maintain an even balance in an oscillating system. Both are considered the most preferred foods and the Miskito say that they never tire of eating them.

Cassava is able to produce large yields in sandy beach soils and can be left in the ground until needed. It can also be eaten in large amounts with little taste rejection. Cassava comprises 50% of subsistence agricultural plantings and approximately 40% of the diet by weight in Tasbapauni. Turtle migration takes place during the rainy season when turtling efforts would be poor at best. Turtles will stay alive for a number of days after being caught, thereby allowing meat to be butchered according to the village's needs rather than dictated by the rate of decomposition. In addition, turtle is the most preferred meat and the most available. Green turtle meat makes up 70% of all meat eaten in Tasbapauni and about 10% of the diet by weight. One from the land, the other from the sea, cassava and turtle help stabilize and dampen fluctuation in labor inputs and subsistence behavior.

The Miskito can regulate the amount and direction of the flow of foodstuffs into the village by the degree of intensity with which the resource base is exploited and through the nature of their food-giving social relationships. The admired Miskito behavioral trait of generosity and the need to have food to acknowledge social obligations act to regulate and more evenly distribute food in the village. I would like to suggest that a close relationship exists between the number of food receivers and the degree of labor-exploitation intensity. This relationship might be depicted as in Fig. 58.

Similarly, the number of activated food-giving social channels appear to increase through either food surplus or food need. Thus a turtleman with 100 pounds of turtle meat may distribute a considerable amount

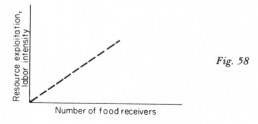

Fig. 58

to reduce his portion to socially acceptable levels, coincident with family needs, while meat will be given to individuals occupying specific priority ranks who may be either too old or sick to acquire meat on their own or may have been unlucky in securing meat that particular day. The more meat that is obtained the wider is its circulation through social relationships and the greater is the equality of meat receipt. This may be shown as in Fig. 59.

One of the major variables associated with cultural adaptations in the subsistence system is distance. The distances involved, frequency of travel, and meat returns from hunting and fishing have been described previously. Distance is largely to be equated with water travel by *dori*, the use of which allows the Miskito to efficiently exploit faraway land and sea resources. In agriculture it was suggested that increasing distance reduces intensity of labor input and yields by lowering effort directed toward weeding and pest control. On the other hand, it was shown that the degree of certainty in obtaining meat may be more of a determining factor in hunting and fishing labor inputs than distance per se. In addition, I would like to suggest that Miskito hunting and fishing efforts become more focused with increasing distance from the village. Close to the village the Miskito can range through various biotopes in search of acceptable animals. The farther they go, however, the greater must be their chances for obtaining meat and the more they concentrate on assured meat-yielding biotopes or species (see Fig. 60).

In the regulation of Miskito subsistence there exists a variable F (amount of food consumption) that must be maintained within tolerable limits so that the population can persist. The F variable can be considered as being the minimum intake of calories, protein, and other needed nutrients for an individual, family, or village to allow subsistence at culturally and biologically prescribed limits. Numerous supporting variables affect F, some of which are environmentally regulated such as the availability and accessibility of foods, temperature as it affects energy expenditure and metabolism, and some of which are regulated by the Mis-

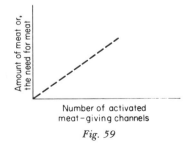

Fig. 59

Fig. 60

kito such as selection of crops, amount of land cultivated, amount of labor input in food getting, hunting and fishing foci, and the extent and degree of reciprocal distribution of food. In addition, the general health status of an individual or population can greatly affect the F variable.

Since sources of foodstuffs are not consistent in occurrence or amount throughout the year, various food-getting means have to be adjusted to coincide with favorable ecological conditions and with each other in order to maintain at least a minimal intake of food. Thus, during the June through August rainy season, when turtling is unproductive and money is difficult to obtain to buy food, fishing, hunting, gathering, and residual crops from the previous year's plantation provide the basis for subsistence (Fig. 61). On the other hand, during the February through May dry season, agricultural supplies are scarce and food-getting efforts are turned to turtling and to the purchase of foodstuffs with money obtained from turtling and shrimping. By focusing on different means of food procurement at different times of the year, an average calorie consumption could be maintained between 2000 and 2800 per day for an adult, despite major ecological changes in the Miskito's environment.

The Miskito's livelihood system adapted early to economic opportunities provided by foreign contact. For the most part, cash resource and labor exchange were carried out away from local systems and above or in addition to satisfaction of traditional Miskito subsistence goals and behavior. With increasing contact and increasing demand for resources, engagement with outside systems began to compete more and more with internal subsistence scheduling and regulation. Many of the Miskito's

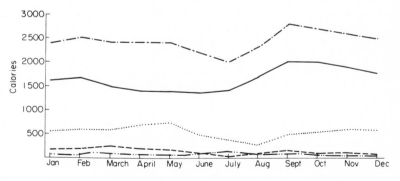

Fig. 61. Approximate contribution of various sources of calories in an average diet at different times of the year. (–·–·): Total calories; (——): Agriculture; (····): Purchased foods; (––): Hunting and fishing; (–··–··): gathering.

reactions and adaptations to outside systems have been and are increasingly maladaptive in that they tend to simplify and degrade ecosystems and decrease social control of access to, and distribution of resources.

The Miskito have participated in trade and wage labor with individuals and companies from the Western world for over 300 years. Economic, social, and subsistence adjustments have been made by the Miskito. Earlier, it was noted that the Miskito are neither peasants nor simply contacted natives. Rather, as suggested by Helms (1969b:328–329), they are a purchase society tied only to the national state by the balance of commercial activities. In order to allow the Miskito to participate in market and wage labor arrangements, a variety of adaptations have taken place in the traditional subsistence system. These adjustments have adapted Miskito subsistence to the need for allowing males the mobility to engage in outside work opportunities. Experiences obtained while working for foreign companies have been fed back into village livelihood patterns. Thus exchange labor agreements have been expanded to include wage labor contracts in the villages. Previously exchanged goods are now sold within the village. In addition, Helms (1970a) has suggested that in order to promote commercial activities Miskito society shifted from patrilocal to matrilocal residence patterns.

As the Miskito became increasingly dependent on outside commercial activities they became more and more dependent on agriculture as the basis of their subsistence system. With the men away, fish and game could not be obtained in large quantities, and women were left to care for agricultural grounds. Homecoming from turtling expeditions to the south or from working for companies was often accompanied by large consumptions of meat. Based on their cultural history of being primarily hunters and fishers, and because meat is considered to be the major strength-giving food, the Miskito continue to regard meat as the determinant of the quality of subsistence. Although based principally on agriculture, Miskito subsistence efforts since the decline of the companies in the mid-twentieth century have been redirected more and more toward specific fish and game animals.

Participation in the market economy determines to a large degree how much food will be available for distribution and consumption. This has caused some disruption of the traditional subsistence system as both seller and buyer are moving into a money-based economy where everyone has to be able to obtain some money in order to buy food, which in the subsistence economy would have been distributed as gifts. The more the village obtains money by devoting labor and attention to particular resources, many of which are selected because of outside market demands, the less meat and other foods there will be for food sharing and the more unequal will become food consumption. Being able to buy food supplies in the village does, however, help to even out daily and seasonal food production for nuclear families caused from disruption of subsistence scheduling. Thus the village market economy has two contradictory functions: it decreases the amount of subsistence food exchanged and smooths out differences in food getting for those who can afford to make purchases (Fig. 62).

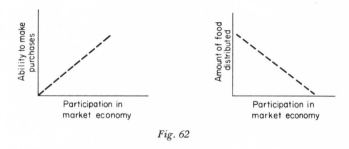

Fig. 62

CAPACITY AND STRESS IN THE SYSTEM

The Miskito's ability to live within their ecosystem without seriously degrading their environment will be determined by their rate of population growth; the capacity of their subsistence system for increased, ecologically balanced production; and the degree of ecological and economic changes resulting from encroaching pioneer Ladino agriculturalists, the market demand for forest and sea resources, and any future resurgence of commercial companies.

Based on an estimated annual growth of 3.3%, the current Miskito population of 35,000 will increase to 48,000 in 10 years and to 66,500 in 20 years. Generally at the limit of their territorial expansion into available waterside resource areas and unable to secure money and foreign goods in the depressed wage labor market, the Miskito are having to support their growing population with local resources and the capability of their subsistence system to accomodate the large proportion of young, nonproductive Miskito who may comprise 50% of a village's population. The functioning of the present subsistence system and its productivity and capacity, as exemplified by the Tasbapauni case, may provide a clue to the future of the Miskito.

Miskito subsistence agriculture has been portrayed as being highly productive on the basis of high yields from relatively low labor inputs.

Hunting and fishing, on the other hand, were shown to be not as productive, but capable of procuring fairly dependable quantities of meat (101,866 pounds of meat in 1 year in Tasbapauni). Turtling was shown to have a higher return of protein per calorie investment and also had a higher chance of success per trip (73% versus 55%) than did hunting. Therefore, even though the Tasbapauni Miskito are on the edge of a still game-rich tropical forest environment, more than 65% of the active adult men concentrate their meat-getting efforts only on turtling.

Calorie production has been in the hands of women who tend the plantations and manufacture coconut oil for home consumption and for sale. Men, however, are increasingly taking part in agricultural activities normally considered women's work, such as planting, weeding, and harvesting. Men are the major procurers of protein, basing their activities on the calorie surplus provided from agriculture.

In order to estimate what capacity the present subsistence system is capable of operating at it is necessary to first discuss some of the problems with the concept of carrying capacity when applied to the Miskito situation. Carrying capacity refers to the maximum level of population at a subsistence level which can be supported in an area with a given technology without "environmental degradation." However, Street (1969) challenged the assumption that technology, cropping patterns, and per capita food consumption are ever qualitatively or quantitatively constant in any given subsistence system. He also remarked that few investigators have attempted to look for environmental degradation and suggested that:

> A pitfall that often traps enquirers into the ecology of shifting agriculturalists is the tendency to become enamored of the object of their scrutiny; the swiddeners appear to be so self-sustaining, so well integrated with their environment, so in harmony with nature that it is hard to believe that they may be damaging their resource base [p. 106].

So far all attempts to measure carrying capacity have been based primarily on how much land is needed to supply a given amount of crops for a given number of people.[6] However, in South America, Carneiro (1960, 1961, 1964) has shown that most tropical forest agricultural systems are capable of producing vegetable foods far in surplus of what is needed by existing populations. It may not be the agricultural potential

---

6. Many geographers and anthropologists have worked with the idea of population carrying capacity among subsistence groups. Allan (1949, 1965), Carneiro (1960), and Conklin (1959), among others, have presented formulas which express the relationship between population, area, and agricultural productivity. Brush (1970) has reviewed the concept and application of carrying capacity for the study of shifting cultivation.

nor the size of the area available for agriculture which influences the size of the population, its mobility, and the area's carrying capacity, but other factors such as the availability of protein rich fish and game resources (Lathrap 1968; Denevan 1966; Carneiro 1970).

Any consideration of carrying capacity should include all sources of food. Even if meat from hunting, let us say, makes up only 5 or 10% of a diet, it may be one of the most important factors in the group's ecological adjustment to the area, and therefore has to be considered along with agricultural yields when assessing carrying capacity. For groups such as the Miskito, Sumu, and Rama of Nicaragua, who are very dependent upon hunting as well as fishing, the land–water area which will support a given population may be many times the size of the area needed for shifting cultivation alone.

The present percentages of calories secured by the Tasbapauni Miskito by different means of food procurement roughly are: agriculture 74%, hunting and fishing 7%, gathering 1%, and participation in the market economy 18%. However, it must be remembered that although hunting and fishing make a low contribution to total calorie intake, they do provide 97% of the animal protein consumed in the village. Given approximately 1,710,000 calories and 19,869 grams of protein as the average amounts consumed every day in Tasbapauni, then some 625 million calories and 7.5 million grams of animal protein are needed for a year's food supply. If we assume that 74% of the calories will come from subsistence agriculture, then, based on measured calorie yields per acre, 232 acres of old and new plantations are sufficient to maintain present calorie levels. However, this figure should perhaps be adjusted to 300 acres because of losses from animal pests and flooding. A rough estimate of the present total area of Tasbapauni subsistence plantations (this year's) would be 275–300 acres.

Using Carneiro's (1960) formula for carrying capacity for agriculture and assuming, for the moment, that all other food sources are capable of being increased proportionately, an estimate of maximum population capacity can be obtained under shifting agriculture alone:

$$P = \frac{[T/(R + Y)] \times Y}{A},$$

where P is the population which may be supported, T is the total arable land in acres, R is the length of fallow in years, Y is the length of cropping period in years, and A is the area of cultivated land required to provide the average individual with the amount of food he ordinarily derives from cultivated plants per year.

Assigning a generous 8-year fallow period, using the figure of 0.3 acres per individual in the village (300 acres per 1000 population), and including an availability of an estimated 7040 acres of the best arable land[7] with a waterside or near a waterside location, the following solution is derived:

$$P = \frac{[7040 \text{ acres}/(8 + 2)] \times 2}{0.3},$$

$P = 4693$

Therefore, a village of almost 5000 persons could live and cultivate in the same area that Tasabapuni villagers do now in contrast to 997 at present (1969 population). But the carrying capacity of the area is not determined by agriculture alone. To maintain the present levels of protein consumption, approximately 3500 green turtles would have to be taken annually for subsistence alone to support a human population of 5000. It is very doubtful that the local green turtle population could withstand an increase of this magnitude. Certainly this degree of exploitation must be considered "environmental degradation," although exactly how much pressure the local green turtle population can stand is not known. Maintaining the present quality and composition of Tasbapauni Miskito dietary patterns in the future will be as closely correlated to what happens to green turtle populations as it will to increasing the area under cultivation to meet population growth.

The Miskito of Tasbapauni and of coastal eastern Nicaragua, have adapted much of their lifeways and subsistence strategies to green turtles. Therefore, what is happening or what will happen to green turtle populations has a great deal to do with the future adaptive patterns of the Miskito.

An important measure of the capabilities of the Miskito's subsistence system is population increase and maintenance of food intake. In order to keep the "F" variable consistent with Miskito ethnodietary standards and biological needs in the face of increasing population, the carrying capacity of a particular area has to be increased proportionately or either environmental or nutritional degradation or both may occur. The Miskito have historically been able to support a rapidly growing population by territorial expansion; by focusing on high yielding agricultural crops such as cassava, bananas, and plantains; and, for the coastal Miskito, by basing much of their protein quest on green turtles. The Miskito

---

7. Estimate made from air photographs and field reconnaissance.

population, however, through internal population growth and external market demand, is subsisting on a much-reduced faunal base. The riverine Miskito, with a denser population than the coastal Miskito, have depleted much of the local faunal resource base and are suffering a decline in animal protein intake. The coastal Miskito also are beginning to experience a general decline in meat intake, due primarily to a reduction in the green turtle population and selling turtles to exogenous systems. Both the riverine and coastal Miskito are turning more to agriculture as the basis for subsistence. In Tasbapauni, the acceptance of dasheen has permitted increased yields in nearby low-lying wet areas, while reducing the need to extend agriculture to more distant sites. Riverine Miskito are including increased amounts of rice and beans, once considered foreign food, to supplement their traditional diet. Thus, carrying capacity has been increased by accepting reduced meat yields, by changing dietary patterns, and by increasing agricultural output.

Tremendous stress is being placed on the Miskito's subsistence system as it is opened up to market economy demands: labor expenditure and resource pressure is being increased, geared not by household composition nor by reciprocal, social food-giving relationships, but regulated by market opportunities, demands, and resource and labor availability. Judging from a limited sample, it appears that labor expenditure in market activities is yielding decreased energy returns compared to subsistence agriculture. The transition from subsistence to market economies is also placing a strain on social relationships and fostering increased pressure on the Miskito's ecosystem for resources over and above subsistence needs.

Since European contact the Miskito have expanded their population and territory within an area which has remained culturally and physically isolated from the national state. Therefore, much of their ability to maintain relatively intensive hunting and fishing activities for such a long time is due to the fact that their habitat was isolated from outside incursion and subject to only Miskito exploitation. Once lumber and banana companies and commercial turtling operations began this situation changed. Coupled with the increased resource-consuming Miskito population, both game habitat and game animals have been reduced. Villages such as Tasbapauni and a handful of other Miskito settlements represent remnants of a way of life once much more widespread in tropical Middle America. Tasbapauni still finds itself in a game-rich environment, but the size of game populations is more and more being influenced by controlling factors outside traditional Miskito society. Even though they have developed a subsistence strategy wherein subsistence risk is reduced and food supply relatively assured, the Miskito are losing the ability to control their subsistence base.

As engagement with the market economy deepens, increased exploitation of ecosystem resources will follow as the Miskito attempt to fulfill cultural needs through market items. Intensified hunting and fishing efforts, for example, are being directed toward animal species already utilized for subsistence (green turtles and white-lipped peccary) as well as toward species traditionally not considered a food or resource article, such as shrimp, jaguars, ocelots, margays, and freshwater otters. Thus, because of the lure of monetary return, Miskito hunting and fishing patterns are intensifying on already utilized animals and are expanding to include a wider range of sought after animals. Increased exploitation has a similar effect on the ecosystem as if the Miskito's population were suddenly enlarged. Therefore, the area's ability to support a population may be decreased because of "cultural mimicry" of an expanded subsistence population.

In general, the maladaptive trends presently evident in Tasbapauni and for all Miskito populations are as follows: (1) increasing simplification of the variety of subsistence; (2) the transition of a stable subsistence economy into an unstable cash economy based on local subsistence resources, with potential adverse nutritional consequences; (3) loss of local autonomy of control of the system to outside systems; (4) a trend toward hypercoherence, or too much coherence with outside systems, which makes the local system too vulnerable to extraneous fluctuations in outside systems (market prices for resources, for example); (5) and the over-all tendency toward reducing the Miskito's general purpose system organized around subsistence for the community's population to a specific purpose system oriented toward differentiated access and attainment of energy sources and materials.

Social responsibilities based on kinship and reciprocity are losing their power to operate as a homeostatic mechanism to regulate subsistence and adapt the Tasbapauni Miskito to their environment. Economic forces instigated from outside systems have set in motion internal oscillating waves within Miskito society and their ecosystem which are tending toward ecological simplification and social atomization. The rules of Miskito society contain little anticipatory power to cope with extraneous economic challenges, nor are feedback mechanisms very efficient in light of increasing monetary dependence to affect restabilization of a threatened system. Margalef (1969) referred to a similar ecological process when he wrote:

> Exploitation is like inflicting a wound upon a heterogeneous organic structure: some tissues or subsystems (more mature) do not regenerate; others (less mature) do. . . . Maintained exploitation keeps the maturity of the exploited system constantly low. . . . More energy goes into fluctuations such as those represented by exploited populations or by populations that are integrated into exploited ecosystems. Extremely mature ecosystems, such as tropical forests, are unable to go back and are totally disrupted

by human exploitation. . . . Human activity decreases maturity and can enhance fluctuations [p. 388].

The Miskito are an extremely adaptable people. They have proven this throughout 300 years of European culture contact. Yet they still have a strong and viable culture based on their subsistence system. Their means of subsistence, like that of many aboriginal and peasant groups in Latin America, are being threatened by forces outside traditional culture. Green turtles have not proven to be so adaptable; they are instinct bound and this is contributing to their decline.

This morning a few turtle dories probably put out from Tasbapauni to the cays. The men went to "look a turtle." The village is waiting for their return.

APPENDIX A

PREVIOUS RESEARCH
ON THE MISKITO COAST

Few geographers have visited the Miskito Coast and fewer have written about it, despite the attraction of the cultural diversity and natural beauty. Parsons (1954, 1955a, b) has dealt with the pine savannas, English settlement, and gold mines of the coast. Radley (1960) wrote a very useful physical geography of the area. Three German geographers have done a *Wanderjahr* along the Nicaraguan and adjacent Honduran shores. Helbig (1965) wrote a book on the land use of the northeastern region of Honduras, focusing a good deal of his attention on the Miskito of that area. Influenced by Helbig, Nuhn (1969) tramped across savanna and mountain trails on the Miskito Coast gathering information for a monograph he is presently working on dealing with the Caribbean lowlands from Belize to Panama. Sapper (1900) wrote a very informative report on his trip to the Río Coco, one of the most densely populated Miskito areas. Another German geographer, Sandner (1964), has written a summary of the historical geography of the Caribbean fringes of Nicaragua, Costa Rica, and Panama.

Geographic studies of adjacent regions and cultures, devoting some attention to the Miskito and to the Miskito Coast, include those by Denevan (1961) on the pine forests of Nueva Segovia, Parsons (1956) on the islands of San Andrés and Providencia, Radell (1969) on the historical geography of western Nicaragua, and Von Hagen (1940) on the Miskito

245

of eastern Honduras. In addition, Conzemius (1929c) has supplied the only detailed description of Corn Island.

Conzemius (1929a, b, 1932, 1938) and Helms (1967, 1969a, b, c, 1970a, b, 1971) are the only anthropologists to work at any length with the Miskito. Conzemius (1932) wrote a brief but valuable ethnography of the Miskito and Sumu Indians. Recently Helms (1971) spent 9 months among the Miskito of the upper reaches of the Río Coco and has written the single-most important report on the Miskito. In the *Handbook of South American Indians*, Kirchhoff (1948) compiled a survey of the Caribbean lowland Indian groups, giving some attention to the Miskito.

The east coast of Nicaragua has been visited by some of the most descriptive traveler–naturalists of any part of the Caribbean. Accounts by Bell (1862, 1899), Belt (1874), Roberts (1827), and Squier (1855) make for exciting reading and recall days when to travel on the Miskito Coast was high adventure.[1] Strangeways (1822), Pim and Seemann (1869), M.W. (1732), and Wickham (1872) wrote very readable descriptions of their travels and experiences which have been extremely helpful in reconstructing past conditions on the coast.

A number of other studies on the coast, its history, and its people are available. The history of the Anglo-Spanish conflict over the coast has been recorded by Floyd (1967). This deals little with the Miskito but does offer valuable historical background, as do works by De Kalb (1893) and Gamez (1939). A recently reissued book by Williams (1969), edited by Dickason, was America's first novel and stems from the author's experiences while shipwrecked on the southern Miskito Coast. Written sometime between 1745–1775, it describes in great detail the natural history of the area and a group of Indians which were probably the Rama. Taylor, an ecologist who was previously with the FAO Mission to Nicaragua, has done work on the vegetation of Nicaragua (1963), including the Atlantic lowlands, and on ecological land use (1959) in the northeast section of the coast. Part of a book by Carr (1953), a zoologist at the University of Florida, related his experiences traveling by dugout canoe through the tropical forests near the area that we studied and provides sensitive insights on the people and countryside. Further investigations which were helpful are by Heath (1950) and Pijoán (1944, 1946) on the Miskito, and by Harrower (1925), Massajoli (1968, 1970), and Mueller (1932) on the Miskito and other Indian groups. Finally, of the available dictionaries and papers on Miskito language, the most useful were those by Heath (1913), and Heath and Marx (1961).

---

1. The latter two books have been reprinted in the University of Florida Gateway Series. Squier's book appears under his pseudonym Saml. A. Bard.

SOME COMMON ANIMALS AND PLANTS

No collections were made for positive identification by specialists. Plants and animals were identified with the aid of various keys and the help of FAO forestry personnel in Puerto Cabezas and a fishery project Peace Corps volunteer in Bluefields. The scientific nomenclature for fish and game animals is fairly certain; the plants, less so.

Common name	Miskito name	Spanish name	Creole name	Scientific name
Mammals				
Agouti	Kiaki	Guatuza	Kiaki	*Dasyprocta punctata*
Anteater, lesser	Wingku-sirpi	Perico lerdo	Small antsbear	*Tamandua tetradactyla*
Anteater, giant	Wingku-tara	Oso hormiguero	Big antsbear	*Myrmecophaga* sp.
Armadillo	Taira	Armado, Armadillo	Jacket man	*Dasypus* sp.
Brocket deer	Snapuka	Venado colorado	Red deer	*Mazama americana*
Brush rabbit	Tiban	Conejo	Rabbit	*Sylvilagus* sp.
Bush dog	Arari	Culumucu	Bush dog	*Speothos*
Coati-mundi	Wistiting	Pisote	Quash or Cuash	*Nasua narica*
Collared peccary	Buksa	Sahino	Pecari	*Pecari tajacu*
Howler monkey	Kungkung	Mono negro	Baboon	*Alouatta* sp.
Jaguar	Limi bulni	Tigre	Tiger	*Felis onca*
Kinkajou	Uyuk	Cuyús	Night walker	*Potos flavus*
Manatee	Palpa	Manatí	Mananti	*Trichechus* sp.
Margay	Limwiata	Peludo, Caucelo	Tigercat	*Felis wiedii*
Mountain lion	Limi pauni	León	Red tiger	*Felis concolor*
Ocelot	Limi buro, Krubu	Tigrillo	Tigercat	*Felis pardalis*
Otter, freshwater	Mamu	Nutria	Water dog	*Lutra*
Paca	Ibina	Guardatinaja, Tepezcuintle	Givenot	*Cuniculus paca*
Racoon	Suksuk	Mapachín	Racoon	*Procyon lotor*
Spider monkey	Ruskika, Urus	Mico	Red monkey	*Ateles geoffroyi*
Tapir	Tilba	Danta	Mountain cow	*Tapirella* sp.
Tree squirrel	Buchong	Ardilla	Squirrel	*Sciurus* spp.
White-faced monkey	Wakling	Cariblanco, Capuchino	White face monkey	*Cebus capucinus* ssp.
White-lipped peccary	Wari	Cancho de monte	Wari	*Tayassu pecari*
White-tailed deer	Sula	Venado	White deer	*Odocoileus virginiana*
Reptiles				
Boa	Waula	Boa	Waula	*Boa* spp.
Caiman	Karas	Lagarto	Alligator	*Caiman*
Crocodile	Tura	Crocodrilo	Crocodile	*Crocodylus*
Green turtle	Wii	Tortuga verde	Turtel	*Chelonia mydas mydas*

Common name	Miskito name	Spanish name	Creole name	Scientific name
Hawksbill turtle	Aksbil	Carey	Aksbil	*Eretmochelys imbricata*
Hicatee turtle	Kuswa	Bocatora ?	Hicatee	*Pseudemys* spp.
Iguana	Kakamuk	Iguana	Guana	*Iguana* spp.
Islu	Islu	Garrobo	?	*Ctenosaura* ?
Birds				
Chachalaca	Wasakla	Chachalaca	Chachalaca	*Ortalis* sp.
Curassow	Kusu	Pavo, Pavo real	Carraso	*Crax* sp.
Macaw	Apawa	Lapa	Maca	*Ara*
Parrot, yellowhead	Rauha	Lora	Parat	*Amazona ochrocephala*
Tinamou	Wangkar	Perdiz	Mountain hen	*Crypturellus* sp.
Toucan	Yamukla	Pícon	Bill bird	*Ramphastus tocard*
Fish				
Barracuda	Barkura	Baracuda	Barracuda	*Sphyraena barracuda*
Catfish	Batsi, Tunki	Bagre, Barbudo	Catfish	*Arius melanopus, Bagre marinus*
Coppermouth, Sea trout	Bilapau	Corbina	Coppermouth	*Cynoscion* sp.
Drum	Pis Pis	Roncador	?	*Menticirrhus martinicensis*
Drummer, Croaker	Susumaya or Walpa-yula	Dormilón, Tambor, Roncador	?	*Micropogon furnieri* ?
Flounder, Flatfish	Daha	Pez hoja	Sand fish	*Botheidae*
Grouper	?	Cabrita	?	*Diplectrum* sp.
Guapote	Sahsing	Guapote	?	*Cichlasoma* spp.
Jack	Krauhi, Kramwi	Jurel	Jack fish	*Caranx hippos*
Jewfish or Junefish	Kunha or Kuha	Mero	June fish	*Epinepaalus* sp.
Mackerel, Spanish	Lasisi ?	Macarela, Sierra	Macrel	*Scomberomorus* sp.
Manta ray, Devilfish	Whipri	Manta	Whiprey or Whipi	*Manta birostris*
Mojarra	Tuba, Klanki	Mojarra	Tuba	*Cichlasoma* spp.
Mullet, Califavor	Kuhkali	Lisa	Califavor	*Mugil* sp.
Porpoise	Lam	Bufeo	Parpas	Order Cetacea
Red snapper	Snapper pauni	Pargo rojo	Red snapper	*Lutjanus* sp.
Rook rook	Rukruk	Roncador	Ruk Ruk	*Scianidae*
Sawfish	Twaina	Pez espada	Sword fish	*Pristis* sp.
Shark	Illili	Tiburón	Shark	Many genera

Common name	Miskito name	Spanish name	Creole name	Scientific name
Sheepshead	Sikuku	Diente mico	Sheepshead	Scaniidae
Snook	Mupi, Kalwas	Róbalo	Snook, Big Bone	Centropomus sp.
Stingray	Kiswa	Raya	Sting ray	Urolophidae or Dasyatidae
Stone bass	Trisu	Palometa	Stone bass	Diapterus rhombeus
Tarpon	Tapam	Sabalo real	Tarpon	Megalops atlanticus
Topminnow, Sardine	Pupu or Popo, Bilam	Sardina	Sardine	Poeciliidae or Characidae
Shellfish–Molluscs–Crustaceans				
Cockles, Clams	Klitu	Almeja de bahía	Carkle	Polymesoda solida
Conchs	Tututia (Kip taya)	Caracol, Cambute	Conk	Strombus sp.
Crab, beach or ghost	Kaiaska	Cangrejo	Crab	Ocypode sp.
Crab, swimmer	Rahti, Solati	Jaiba, Cangrejo	Sea crab	Callinectes sp.
Crab, land or blue	Kaisni	Cangrejo	Blue crab	Cardisoma guanhumi
Fiddler or Mangrove crab	Kangli	Cangrejo de mangle	Crab	Uca sp.
Lobster	Wahsi tara or Wastara	Langosta	Lobster	Panulirus argus
Oyster	Sita	Ostra or Ostión	Oysta	Crassostrea rhizophora
Shrimp	Wahsi sirpi	Chacalín, Camarón blanco	Shrimps	Penaeus schmitti
White, Pink	Wahsi sirpi	Camarón rosado	"	Penaeus duorarum and P. aztecus
Hard head or sea bob	Wahsi sirpi	Chacalín	"	Xiphopeneus kroyeri
Shrimp, river or freshwater	Wahsi krukru or krukru	Camarón or Langostina	Fresh water shrimp	Macrobrachium sp.
Surf clams	Ahi	Almeja	Ahi	Donax sp.
Tulip shell	?	Caracol	?	Fasciolaria tulipa
Whelks, top shell	"Wilks"	Caracol	Wilks	Tegula sp.
Cultivated plants				
Anatto	Aulala	Achiote	Nata	Bixa orellana
Avocado	Sikya	Aguacate	Butter pear	Persea gratissima
Banana	Siksa	Banana	Banana	Musa spp.
Beans	Bins	Frijoles	Beans	Phaseolus vulgarius
Breadfruit	Bredput	Fruta de pan	Breadfruit	Artocarpus communis
Calabash tree	Kahmi	Jícaro	Gourd tree	Crescentia cujete
Cashew	Kasau	Marañon	Cashew	Anacardium occidentale
Cassava, sweet	Yauhra	Yuca	Casadda	Manihot dulcis

Common name	Miskito name	Spanish name	Creole name	Scientific name
Chili pepper	Kuma	Chile	Pepper	*Capsicum* sp.
Coconut palm	Kuku	Coco	Coconut	*Cocos nucifera*
Corn, maize	Aya	Maíz	Corn	*Zea mays*
Dasheen	Dasheen	Malanga, Ñampí	Dasheen	*Colocasia esculenta*
Duswa	Duswa	Tiquisque	Coco	*Xanthosoma* spp.
Grapefruit	Sadik, Shadik	Tronja	Grapefruit	*Citrus paradisii*
Lemon	Leimus, damni	Limón	Sweet lemon	*Citrus limon*
Lime	Leimus	Lima	Lime	*Citrus aurantifolia*
Mango	Mango, Mankro	Mango	Mango	*Mangifera indica*
Orange	Anris	Naranja	Aringe	*Citrus sinensis*
Papaya	Twas	Papaya	Papaya	*Carica papaya*
Pejibaye	Supa	Pejibaye	Supa	*Guilielma utilis*
Pineapple	Pihtu	Piña	Pine	*Ananas comosus*
Plantain	Platu	Plátano	Plantin	*Musa paradisiaca*
Plas	Plas (flas)	Guineo	Costo	*Musa sapientum* ?
Pumpkin, squash	Iwa	Calabaza	Punkin	*Cucurbita pepo*
Rice	Rais	Arroz	Rice	*Oryza sativa*
Rose apple	Rose apple	?	Rose apple	*Eugenia jambos*
Soursop	Dwarsop, Puntara	Guanábana	Sawasop	*Annona americana*
Sugarcane	Kayu	Caña de azucar	Cane	*Saccharum officiarum*
Sweet potato	Tawa	Batatas	Sweet potato	*Ipomaea batatas*
Tamarind	Slim, Tamrand	Tamarindo	Tamrand	*Tamarindus indica*
Yam	Tamus, Yamus, Paisawa	Name	Yampi	*Dioscorea* spp.
Wild and feral plants				
Bihu	Bihu	?	Bihu	*Melastomo* ?
Breadnut	Tisba	O-joche	?	*Helicostylis ojoche*
Bribri	Bribri	Guajiniquil	?	?
Cacao	Kakau	Cacao	Cacao	*Theobroma cacao*
Coco plum	Tawa	Hicaco	Cocaplum	*Chrysobalanus icaco*
Chiny root	?	?	?	?
Granadilla	Drap	Granadilla	Granadilli	*Passiflora* sp.

Common name	Miskito name	Spanish name	Creole name	Scientific name
Guava	Sigra	Guava	Guava	*Psidium guajava*
Hog plum	Pahara	Jobo	Hog plum	*Spondias lutea*
Ibo	Ibo	Almendro	Ibo	*Coumarouna*
Kisu	Kisu	Coyolillo	Comombo boy	*Astroncarium ?*
Lemon grass	?	?	Fever grass	*?*
Locust	Laua	Guapinol	Locus	*Hymenaea courbaril*
Mamey	Kuri	Sapote	Mami apple	*Lucuma mammosa*
Monkey apple	Kuramaira	Urraco	Monkey apple	*Moquilea platypus*
Nance	Krabo	Nance	Krabo	*Byrsonima crassifolia*
Sapodilla	Iban	Níspero	Sapodila	*Sapota zapotilla*
Pingwing	Pingwing	?	Pingwing	*?*
Sea grape	Waham	Uva de la playa	Grape	*Coccoloba uvifera*

Commonly used trees and miscellaneous plants mentioned in text

Common name	Miskito name	Spanish name	Creole name	Scientific name
Balsa	Mimi or Mhimhi, Puhlak	Balsa	Balsa	*Ochroma*
Button mangrove	?	?	Buttonwood	*Conocarpus erectus*
Caribbean pine	Auas	Ocote	Pine	*Pinus caribaea*
Cecropia	Plang	Guarumo	Trumpet	*Cecropia peltata*
Cedar	Yalam	Cedro real	Cida	*Cedrela*
Cedro macho	Saba	Cedro macho	Saba	*Carapa guaianensis*
Chicle	Iban and others	Chicle	Chiclet	*Various genera*
Cohune, Corozo	Silal	Corozo	Cohone	*Attalea cohune*
Hone	Ohume	?	Hone	*Corozo oleifera*
Ironwood	?	Comenegro	?	*Dialium guianense*
Kira	Kira bip pata	Guáismo ternero	Brednut ?	*Guazuma ulmifolia*
Mahogany	Yulu	Caoba	Mahagani	*Swietenia macrophylla*
Papta	Papta	?	Papta	*Acoelorrhaphe wrightii*
Red mangrove	Laulu pauni	Mangle	Red mangro	*Rhizophora mangle*
Rubber tree	Tasa dusa	Hule	Rubba tree	*Castilla*
Sandpaper tree	Yahal	Chaparro	?	*Curatella americana*
Sani	Sani dusa	Majagua	Moho	*Hibiscus tiliaceus*
Santa María	Krasa	Santa María	Santa María	*Calophyllum brasiliense*

Common name	Miskito name	Spanish name	Creole name	Scientific name
Silk-cotton tree	Sisin	Ceiba	Cotten tree	*Ceiba pentandra*
Tunu	Tunu	Tunu	Tunu	*Poulsenia armata*
White mangrove	Laulu pihini, Pahtang	Mangle blanco	White mangro	*Laguncularia racemosa*
Beach morning glory	?	?	May wit	*Ipomoea pes-caprae*
?	Sinimint dusa	Canel	Cinnamon tree	*Nectandra globosa*
?	Ihinsa	Guayavo negro	?	*Terminalia amazonia*
?	Kira pauni	Guásimo colorado	Guásimo	*Luehea seemannii*
?	?	Coralillo ?	?	*Erthrina glauca*

253

APPENDIX C

ITEMS, WAGES, AND PRICES
IN THE MARKET ECONOMY

Activity–job–item	Wage–price (in *córdobas*: 1 *córdoba* = $ .14)		
	1969		1971
Turtle products			
Fresh calipee	2.00	per pound in Tasbapauni	2.50
Dried calipee	5.00	per pound in Tasbapauni	5.50
	7.00	per pound in Bluefields	7.00
Hawksbill shell second class	12.00	per pound in Tasbapauni ⎱	15.00
first class	14.00	per pound in Tasbapauni ⎰	
	21–22.00	per pound in Bluefields	23–25.00
Dori work			
Cutting and fitting	300–500.00	plus materials	
Sail making	25.00		
Hunting and fishing equipment			
Turtle "peg"	5.00		6–7.00
Harpoon points	1.00	a barb	
Shrimp net	100.00	labor and materials	115–120.00
Agricultural work			
Cutting–clearing	7.00	a day and food (axe work)	8.00
	5.00	a day and food (machete work)	7.00
Planting rice	2.00	a day and food—woman	2.50
	3.00	a day and food—man	3.00
Replanting rice	5.00	a day with food	
Harvesting rice	5.00	100 pounds	
		or	
	3.50	100 pounds and food	
Picking up debris	3.00	a day and food—woman	
Weeding	3.00	a day and food	5.00
Animals			
Green turtle meat	.50	per pound	.75
Fish	.25	per pound—"grab"	
Jaguar and ocelot skins	750–900.00	in Bluefields	

254

Activity–job–item	Wage–price (in *córdobas*: 1 *córdoba* = $ .14)		
	1969		1971
Margay skin	30.00	in Bluefields	
River otter skin	15–25.00	in Bluefields	
Shrimp	10–14.00	30 pound can (depending on size of shrimp)	
Live turtle	35–70.00	in Bluefields (depending on size)	60.00, Bluefields turtle company
Deer meat, and white-lipped peccary meat	.75	per pound	.90
Manatee meat	.80	per pound	1.00
Beef	1.50	per pound	
Dried shrimp	1.75–2.00	per pound	3.00
Pigs	125–150.00	a large one—Bluefields buyers	
Chickens	4–5.00	apiece—Bluefields buyers	
Coconuts	15.00	per 100 nuts	
Coconut oil	9.00	1 gallon (40 nuts) in Tasbapauni	10.00
	10.00	1 gallon (40 nuts) in Bluefields	12–13.00
Rice	50.00	100 pounds hulled, in Bluefields	
Cassava	.20	pound or "grab"	
Wild cane, Corn Island	20.00	100 pounds	
Jobs in village			
Lay pastor	100.00	monthly plus	
School teacher	400–500.00	monthly	
Boat crew/boat owner	by trip		
Store keeper	—		
Carpenter—wood cutter	by item		12–20.00 per day
Jobs outside of village			
Shrimp boats	by job		
Mines	by job		
Cocal coconut plantation	by job		
Lumbermill in Bluefields	by job		

STORE GOODS AND PURCHASES

MOST COMMONLY SOLD ITEMS
In Córdobas and Amounts (One Córdoba = $ .14)

	1971	1969	
Flour	.90	.80	pound
Salt		.75	pound US; .60 pound Nicaraguan
Sugar		.90	pound
Baking powder		4.00	tin, usually sold in .15, .20, .25 amounts
Yeast		.25	2 teaspoonfuls
Coffee		.25	ounce, C$4.00 pound
Beans		1.00	pound
Rice	1.10	.80	pound
Coconut oil		1.00	pint, almost always sold in .25, .50, .75 amounts
Bread		.25	bun (approximately 3 ounces)
Spices		.20	cloves, peppercorns, cinnamon, vanilla, *achiote* (Annatto) and nutmeg are sold in quantities of .10–.40 for special occasions
Oatmeal		4.00	tin, usually sold in .10, .15, .20 amounts for the sick
Cocoa		5.50	tin, usually sold in .15, .20, .25 amounts
Candies		.05	apiece
Soda crackers		.05	apiece
Kerosene		.50	pint
Soap		1.00	a bar
Chiripa		.25	ounce, drink concentrate
Cigarettes		.25	3, a pack is 1.25–2.25
Fish line		.10	foot
Shotgun shell	2.25	2.00	1
.22 shell		1.00	3
Aspirin	.15(1)	.15	2
BandAid		.15	1, store owner cleans cuts and puts disinfectant on for same price
Vicks		1.00	small tin

Purchases of Store Goods: Amounts and Types
(Two one-hour periods)

In córdobas or amounts (c = child; m = man; w = woman)

February 8, 1969		10:00–11:00 a.m.
c	6 pounds	flour
c	.25	salt
c	.25	soap
w	.25	salt
c	.25	coconut oil
c	.25	vinegar
c	.20	paper and envelope
c	.25	yeast
c	2 pounds	flour
c	1.50	worth of coarse pink cotton cloth
c	½ pound	rice
c	.25	coconut oil
c	.50	vinegar
c	.10	grease
m	1 pound	nails
c	.15	black peppercorns
c	.25	coconut oil
c	.25	salt
c	.25	¼ of an onion
c	.25	sugar
c	.25	coffee
c	.25	sugar
c	.50	coconut oil
c	½ pound	salt
c	.25	coconut oil
c	2 pounds	flour
c	.20	yeast
c	1	bar of soap
c	3	coconuts
c	.50	margarine
c	.25	cinnamon
c	.25	coconut oil
c	3 pounds	flour
c	.25	baking powder
c	1 pound	sugar
c	1 pound	rice
c	2	sulfa pills
c	.10	hard candy
c	.25	coconut oil
c	1	bar of soap
February 8, 1969		5:00–6:00 p.m.
c	.50	sugar
c	.50	sugar
c	.05	candy
c	.25	kerosene
c	.25	coconut oil
c	½ pound	sugar

February 8, 1969   5:00–6:00 p.m.

c	2	cigarettes
c	4	breads
c	.20	candy
c	.20	crackers
c	.25	kerosene
c	½ pound	sugar
c	.50	margarine
c	3	fever pills for baby
c	1	writing paper
c	.25	cocoa
c	1	nipple for baby bottle
m	1½ pounds	sugar
m	.15	baking powder
m	2 pounds	flour
c	.25	sugar
c	1 pound	sugar
c	½ pound	salt
c	2	notebooks for school
c	1 pound	flour
c	½ pound	sugar
c	.25	baking powder
c	3 pounds	sugar
c	½ pound	sugar
c	2	fever pills
c	1.00	salt
m	2	flints
c	1 yard	line
m	2	bread
c	1	Presto (instant) coffee, for author and family
c	4	bread
c	.50	kerosene
c	4	bread
c	½ pound	sugar
c	.25	kerosene
c	.25	kerosene
c	.25	kerosene
c	.25	sugar
c	2	cigarettes
c	.25	coffee
c	4	bread
c	1	"bad belly" pill
c	1	aspirin
c	.35	"sena," used for a laxative
c	.25	kerosene
c	.50	kerosene
w	2	bread
c	½ pound	sugar
c	.25	vaseline
c	2	cigarettes
m	3	bread
w	2	aspirin

GLOSSARY

Agouti: A small rodent (*Dasyprocta punctata*) called *kiaki* by the Miskito. About the size of a large jack-rabbit, reddish brown to dark brown in color, the agouti usually lives in dense undergrowth.

Biotope: An ecological unit where primary habitat conditions, and the fauna and flora adapted to them are uniform.

Breadkind: A Creole word used by the coastal Miskito to include many starchy root, tuber, and tree crops, such as cassava, dasheen, breadfruit, and bananas.

Calipee: The cartilaginous amber-colored substance obtained from inside the lower and upper shells of the green turtle. Calipee is exported by Ladino, Chinese, and Creole merchants to the United States and to England where it is used as the basis for green turtle soup.

Cay: A small island. Called *ki* in Miskito it can be anything from a coral island to a small patch of vegetation rising above a flooded area.

Crawl: An enclosure usually made of coral rocks or mangrove branches in shallow water used to contain green turtles. The Miskito refer to the enclosures as *klars*.

Creole English: An English-based Caribbean dialect spoken by West Indian Negroes, Cayman Islanders, Corn Islanders, other Caribbean peoples, and some Miskito on the east coast of Nicaragua.

Dasheen: A tuber-yielding plant (*Colocasia esculenta*) with leaves similar to *duswa* in appearance.

Duswa: The Miskito name for *Xanthosoma*, an American relative of the Polynesian taro or dasheen. *Duswa* is called "coco" in Creole and "tisquisque" in Spanish.

Dori: A keeled dugout canoe used for sailing and turtle fishing by the Miskito. Called a "cayuca" in eastern Costa Rica.

Green Turtle: A large (150–400 pound) herbivorous sea turtle (*Chelonia mydas*) occurring in large numbers off the coast of eastern Nicaragua.

Ground Provisions: A Creole word used for starchy, belowground roots and tubers, such as cassava and *duswa*.

Haulover: A narrow neck of land separating two bodies of water, such as a lagoon and the sea or two lagoons, where a *dori* can be hauled over.

Hawksbill: A medium size (75–150 pound) sea turtle (*Eretmochelys imbricata*) exploited for its translucent, mottled shell, better known as "tortoise shell."

Hicatee: A small (8–20 pound) freshwater turtle (*Pseudemys*) called *kuswa* by the Miskito. This turtle is easily caught and is an important food item in many parts of the coast.

259

Ladino: A term applied to Spanish-speaking Central Americans. The Miskito and Creoles refer to Nicaraguan Ladinos as "Spaniards" or "Sponmon."

Paca: A good-eating nocturnal rodent (*Cuniculus paca*) called *ibina* in Miskito. This is one of the best New World wild meats.

Pejibaye: A species of thorn-ringed palm (*Guilielma utilis*) which yields bunches of yellow-orange, oil-rich fruits twice annually, the major harvest occurring from September through October. The Miskito call this palm *supa*.

Pitpan: A keel-less dugout canoe, which is poled and paddled on rivers and quiet lagoons.

Plantation: The commonly used term for a swidden plot; called *insla* in Miskito and "milpa" in Spanish. The Miskito refer to an old plantation ("rastrojo" in Spanish) as a *prata*.

Plas: A short, thick, square-shaped fruit (*Musa sapientum*) similar to a plantain and known as "cuadrado" or "guineo" in Spanish. Supposedly the Miskito word *plas* was derived from "flask," the square bottles once used by English buccaneers for beverages.

Tas: An area measurement used by the Miskito, 50 yards on a side; 4 *tas* equal 1 manzana or 1000 square yards. Two *tas* (45,000 square feet) roughly equal 1 acre (43,560 square feet).

Wari: The Miskito name for white-lipped peccary (*Tayassu pecari*), a large wild pig weighing up to 80 pounds, which travels in droves of 100–200 individuals.

REFERENCES

Adams, Richard N.
1956   Cultural components of Central America, *American Anthropologist* **58**, 881–907.
Alexander, E.A.
1965   "A Soil Survey of North East Nicaragua." F.A.O. Interim Report No. 8, Puerto Cabezas, Nicaragua. (Processed.)
Allan, William
1949   *Studies in African land usage in northern Rhodesia.* Rhodes-Livingstone Papers, No. 15. London: Oxford Univ. Press.
1965   *The African husbandman.* New York: Barnes and Noble.
Alland, Alexander Jr.
1970   *Adaptation in cultural evolution: An approach to medical anthropology.* New York: Columbia Univ. Press.
Allee, W.C. and Karl P. Schmidt
1951   *Ecological animal geography,* 2nd ed. New York: Wiley.
Bates, Marston
1960   *The forest and the sea.* Signet Science Library, New York: New American Library.
Beals, Alan R.
1964   Food is to eat: The nature of subsistence activity, *American Anthropologist* **66**, pp. 134–136.
Bell, Charles N.
1862   Remarks on the Mosquito Territory, its climate, people, productions. . ., *Journal of the Royal Geographical Society,* **32**, 242–268.
1899   *Tangweera: Life and adventures among gentle savages.* London.
Belt, Thomas
1874   *The Naturalist in Nicaragua.* London.
Bennett, Charles F.
1959   "The Procurement and Utilization of Native Animals by Three Different Culture Groups in the Republic of Panama: A Study of an Aspect of Tropical Biogeography." Unpublished Ph.D. dissertation. Dept. of Geography, Univ. California, Los Angeles.
1962   The Bayano Cuna Indians, Panama: An ecological study of livelihood and diet, *Annals of the Association of American Geographers* **52**, 32–50.
1967   A review of ecological research in Middle America, *Latin American Research Review* **2** No. 3, 3–27.

261

1968   *Human influences on the zoogeography of Panama.* Ibero–Americana, No. 51, Berkeley: Univ. of California Press.
1971   Animal geography in Latin America. In *Geographic research on Latin America,* edited by B. Lentnek, Robert L. Carmin, and T.L. Martinson. Pp. 33–40. Muncie: Ball State Univ.

Blumenstock, D.I.
1958   Distribution and characteristics of tropical climates, *Proceedings of the Ninth Pacific Congress,* Symposium on climate, vegetation and rational use of the humid tropics, **20,** pp. 3–24. Bangkok: Dept. of Science.

Borzek, Josef M.
1966   Food as an essential: Experimental studies on behavioral fitness. In *Food and civilization: A symposium,* edited by S.M. Farber, N.L. Wilson, and R.H.L. Wilson. Pp. 29–60. Springfield, Illinois: Charles C. Thomas.

Bowes, A. and C.F. Church
1966   *Food values of portions commonly used.* 10th ed. Philadelphia: Lippincott.

Bradley, Alice V.
1956   *Tables in food values.* Peoria, Illinois: Bennett.

Brookfield, H.C.
1968   New directions in the study of agricultural systems in tropical areas. In *Evolution and environment,* edited by E.T. Drake. Pp. 413–439. New Haven: Yale Univ. Press.

Brush, Stephen B.
1970   "The concept of carrying capacity for systems of shifting cultivation." Unpublished M.A. thesis. Dept. of Anthropology, Univ. of Wisconsin.

Carneiro, Robert L.
1957   "Subsistence and social structure: An ecological study of the Kuikuru Indians." Unpublished Ph.D. dissertation. Dept. of Anthropology, Univ. of Michigan.
1960   Slash-and-burn agriculture: A closer look at its implications for settlement patterns. In *Men and cultures: Selected papers of the Fifth International Congress of Anthropological and Ethnological Sciences,* edited by F.C. Wallace. Pp. 229–234. Philadelphia: Univ. of Pennsylvania Press.
1961   Slash-and-burn cultivation among the Kuikuru and its implications for cultural development in the Amazon Basin. *The evolution of horticultural systems in native South America, causes and consequences: A symposium,* edited by J. Wilbert. Pp. 47–67. Caracas: Sociedad de Ciencias Naturales La Salle.
1964   Shifting cultivation among the Amahuaca of eastern Peru, *Völkerkundliche Abhandlungen* **1,** 9–18.
1970   The transition from hunting to horticulture in the Amazon Basin, *Proceedings VIIIth Congress of Anthropological and Ethnological Sciences, Tokyo and Kyota, 1968* **3** Ethnology and Archaeology, 244–248. Tokyo: Sci. Council of Japan.

Carr, Archie
1950   Outline for a classification of animal habitats in Honduras, *Bulletin American Museum of Natural History* **94,** 563–594.
1953   *High jungles and low.* Gainesville: Univ. of Florida Press.
1956   *The windward road: Adventures of a naturalist on remote Caribbean shores.* New York: Knopf.
1967   *So excellent a fishe: A natural history of sea turtles.* New York: The Natural History Press.
1969   Sea turtle resources of the Caribbean and Gulf of Mexico, *International Union for the Conservation of Nature* **2,** No. 10, 74–83.
1972   Great reptiles, great enigmas, *Audubon* **74,** 24–35.

Chagnon, Napoleon A.
  1968   *Yąnomamö: The fierce people.* New York: Holt, Rinehart and Winston.
Chayanov, Alexander V.
  1966   *The theory of peasant economy,* (edited by D. Thorner, B. Kerblay, and R.E.F.
         Smith). The American Economic Association. Homewood, Illinois: Richard D.
         Irvin.
Chisholm, Michael
  1962   *Rural settlement and land use: An essay in location.* London: Hutchinson.
Clarke, William C.
  1971   *Place and people: An ecology of a New Guinean community.* Berkeley: Univ. of Califor-
         nia Press.
Coe, Michael D. and Kent V. Flannery
  1967   *Early cultures and human ecology in south coastal Guatemala. Smithsonian Contributions
         to Anthropology* **3.** Washington, DC: Smithsonian Institution.
Collier, Albert
  1964   The American Mediterranean. In *Handbook of Middle American Indians,* edited
         by R.C. West. Vol. 1, pp. 122–142. Austin: Univ. of Texas Press.
Conklin, Harold C.
  1957   *Hanunóo agriculture: A report on an integral system of shifting cultivation in the Philip-
         pines.* FAO Forestry Development Paper, No. 12. Rome: Food and Agricultural
         Organization of the United Nations.
  1959   Population-land balance under systems of tropical forest agriculture, *Proceedings
         of the Ninth Pacific Science Congress of the Pacific Science Association,* 1957, **7,** p.
         63. Bangkok: Secretariat, Ninth Pacific Science Congress.
  1963   *The study of shifting cultivation.* Washington, DC: Pan American Union, Secretaría
         General, Organización de Los Estados Americanos.
Conzemius, Eduard
  1929a Notes on the Miskito and Sumu languages of eastern Nicaragua and Honduras,
         International Journal of American Linguistics **5,** 57–115.
  1929b Die Rama-Indianer von Nicaragua, *Zeitschrift für Ethnologie* **59,** 291–362.
  1929c Les Iles Corn du Nicaragua, *La Géographie* **52,** pp. 346–362.
  1932   *Ethnographical survey of the Miskito and Sumu Indians of Honduras and Nicaragua.*
         Smithsonian Institution, US American Ethnology Bulletin 106, Washington, DC:
         Smithsonian Institution.
  1938   Les tribus indiennes de la côte des Mosquitos, *Anthropos* **33,** 910–942.
Cotheal, Alexander I.
  1848   A grammatical sketch of the language spoken by the Indians of the Mosquito
         Shore, *Transactions of the American Ethnological Society* **2,** 235–264.
Craig, Alan K.
  1966   *Geography of fishing in British Honduras and adjacent coastal areas.* Coastal Studies
         Institute Technical Report No. 28. Baton Rouge: Coastal Studies Institute,
         Louisiana State Univ.
Cry, George W.
  1965   *Tropical cyclones of the North Atlantic ocean.* Technical Paper 55, US Weather
         Bureau. Washington, DC: Dept. of Commerce.
Dampier, William
  1968ed.*A new voyage round the world.* New York: Dover Publications.
Dansereau, Pierre
  1966   Ecological impact and human ecology. In *Future environments of North America,*
         edited by F. Fraser and J.P. Milton. Pp. 425–462. Garden City: The Natural
         History Press.

Davis, Arthur Powell
  1902   Hydrography of the American Isthmus, *Twentysecond Annual Report of the U.S. Geological Survey to the Secretary of the Interior,* 1900–1901. Part IV, Hydrography, pp. 507–630. Washington.

De Kalb, Courtenay
  1893   Nicaragua: Studies on the Mosquito shore in 1892, *Journal of the American Geographical Society of New York* **25**, 236–288.

De Lussan, Raveneau
  1930   *Raveneau de Lussan. Buccaneer of the Spanish Main and early French filibuster of the Pacific: A translation into English of his journal of a voyage into the South Seas in 1684 and the following years with the filibusters,* edited by M.E. Wilbur. Cleveland: Clark.

Denevan, William M.
  1961   *The upland pine forests of Nicaragua: A study in cultural plant geography.* Univ. of California Publications in Geography, **12**, No. 4, 251–320. Berkeley: Univ. of California Press.
  1966   A cultural–ecological view of former aboriginal settlement in the Amazon Basin, *The Professional Geographer* **18**, 346–351.
  1971   Campa subsistence in the Gran Pajonal, eastern Peru, *Geographical Review* **61**, 496–518.

Doran, Edwin B.
  1953   A physical and cultural geography of the Cayman Islands. Unpublished Ph.D. thesis, Dept. of Geography, Univ. of California, Berkeley.

Exquemelin, Alexandre Olivier
  1686   *Histoire des Aventuries qui se sont signalez dans les Indes.* 2 Vols. Paris.
  1856   *The history of the buccaneers of America.* Boston: Sanborn, Carter and Bazin.

FAO
  1950   *Report of the FAO mission for Nicaragua.* Washington, DC.

Fernández, León
  1881–1907   *Colección de documentos para la historia de Costa Rica.* 10 Vols. San José, Costa Rica.

Firth, Rosemary
  1943   *Housekeeping among Malay peasants.* London: The Athlone Press.

Floyd, Troy S.
  1967   *The Anglo–Spanish struggle for Mosquitia.* Albuquerque: Univ. of New Mexico Press.

Gamez, José D.
  1939   *Historia de la Costa de Mosquito (to 1894).* Managua.

Garnier, B.J.
  1961   Mapping the humid tropics: Climatic criteria, *Geographical Review* **51**, 339–346.

Geertz, Clifford
  1963   *Agricultural involution.* Berkeley; Univ. of California Press.

Goldsmith, Grace A.
  1966   Activity, climate, and food, *Food and civilization: A symposium,* edited by S.M. Farber, N.L. Wilson, and R.H.L. Wilson. Pp. 157–179. Springfield, Illinois: Thomas.

Greenberg, Joseph
  1960   The general classification of Central and South American languages. In *Men and Cultures,* edited by A.F.C. Wallace. Pp. 791–794. Philadelphia: Univ. of Pennsylvania Press.

Greenfield, Sydney
  1965   More on the study of subsistence activities, *American Anthropologist* **67**, 737–744.

Grigg, David
  1970   *The harsh lands.* New York: St. Martin's Press.
Hagen, E.
  1961   Analytical models in the study of social systems, *American Journal of Sociology* **67**, 144–151.
Hall, P.
  1966   *Von Thünen's isolated state.* English edition of *Der isolierte Staat.* Oxford: Pergamon.
Harris, David R.
  1971   The ecology of swidden cultivation in the upper Orinoco rain forest, Venezuela, *Geographical Review* **61**, 475–495.
  1972   The origins of agriculture in the tropics, *American Scientist* **60**, 180–193.
Harris, Marvin
  n.d.   "Cultural Energy." Unpublished paper, cited by Rappaport (1968:261–262).
Harrower, David E.
  1925   Rama, Mosquito and Sumu, of Nicaragua, *Indian Notes* **2**, 44–48.
Harvey, David
  1970   *Explanation in geography.* New York: St. Martin's Press.
Heath, G.R.
  1913   Notes on Miskito grammar and on other Indian languages of eastern Nicaragua, *American Anthropologist* **15**, 48–62.
  1950   Miskito glossary, with ethnographic commentary, *International Journal of American Linguistics* **16**, 20–34.
  n.d.   "Medicinal plants." Unpublished manuscript, 15 pp.
Heath, G.R. and W.G. Marx
  1961   *Diccionario Miskito-Español, Español-Miskito.* 2nd ed. Tegucigalpa, Honduras: Imprenta Calderon.
Hegsted, D.M.
  1954   *Balance studies with macro-elements in methods for evaluation of nutritional adequacy and status.* Washington, DC: National Research Council.
Helbig, Karl M.
  1959   *Die Landschaften von Nordost-Honduras.* Hamburg: Hermann Haack.
  1965   *Areas y paisajes del nordeste de Honduras.* Tegucigalpa: Banco Central de Honduras.
Helms, Mary W.
  1967   "Frontier society: Life in a Miskito village in eastern Nicaragua." Unpublished Ph.D. dissertation. Dept. of Anthropology, The Univ. of Michigan.
  1969a   The cultural ecology of a colonial tribe, *Ethnology* **8**, 76–84.
  1969b   The purchase society: Adaptation to economic frontiers, *Anthropological Quarterly* **42**, No. 4, 325–342.
  1969c   Peasants and purchasers: Preliminary thoughts on a differentiation of intermediate societies. In *Peasants in the Modern World*, edited by P. Bock. Pp. 69–74. Albuquerque: Univ. of New Mexico Press.
  1970a   Matrilocality and the maintenance of ethnic identity: The Miskito of eastern Nicaragua and Honduras, *Proceedings of the XXXVIII Congress of Americanists*, Munich, 1968, 459–464.
  1970b   Matrilocality, social solidarity, and culture contact: Three case histories, *Southwestern Journal of Anthropology* **26**, 197–212.
  1971   *Asang: Adaptations to culture contact in a Miskito community.* Gainesville: Univ. of Florida Press.
Henderson, Captain George
  1811   *An account of the British settlement of Honduras, and sketches of the manners and customs of the Mosquito Indians.* 2nd ed. London.

266   References

Henry, Jules
    1951   The economics of Pilagá food distribution, *American Anthropologist* **53**, 187–219.
Hipsley, Eben H. and Nancy E. Kirk
    1965   *Studies of dietary intake and the expenditure of energy by New Guineans.* South Pacific Commission Technical Paper No. 147. Nouméa, New Caledonia: South Pacific Commission.
Hodgson, Colonel Robert
    1822(ed)   *Some account of the Mosquito Territory; Contained in a memoir written in 1757, etc., now first published from the ms. of the late Colonel Robert Hodgson.* Edinburgh.
Hymes, Dell (ed)
    1971   *Pidginization and creolization of languages.* Cambridge: The Univ. Press.
INCAP
    1966   "Recommendaciones nutricionales diarias para las poblaciones de Centro America y Panama." Publicaciones Cientificas del Instituto de Nutrición de Centro América y Panamá, Recopilación No. 5, 75–76.
Johnson, Frederick
    1940   The linguistic map of Mexico and Central America. In *The Maya and Their Neighbors,* edited, pp. 88–125. New York: D. Appleton Century.
Kemble, Stephen
    1884   Report on the Mosquito country, *New York Historical Society Collections* **17**, 419–431.
Kirchhoff, Paul
    1948   The Caribbean lowland tribes: the Mosquito, Sumu, Paya and Jicaque. In *Handbook of South American Indians,* edited by J. Steward. US Bureau of American Ethnology Bulletin No. 143, Vol. **4**, 219–229.
Klingel, Gilbert C.
    1961   *The ocean island (Inagua).* A Doubleday Anchor Book, Natural History Library. Garden City, New York: American Museum of Natural History.
Küchler, A.W.
    1961   Mapping the humid tropics: Vegetative criteria, *Geographical Review* **51**, 346–347.
Lade, Robert
    1744   *Voyages du capitaine Robert Lade en différentes parties de l'Afrique, de l'Asie et de l'Amérique.* 2 Vol. Paris.
Lasker, Gabriel N.
    1969   Human biological adaptability: The ecological approach in physical anthropology, *Science* **166**, 1480–1486.
Lathrap, Donald W.
    1968   The 'hunting' economies of the tropical forest zone of South America: An attempt at historical perspective. In *Man the hunter,* edited by R.B. Lee and I. Devore. Pp. 23–29. Chicago: Aldine.
Leeds, Anthony
    1961   Yaruro incipient tropical forest horticulture: Possibilities and limits. In *The evolution of horticultural systems in native South America, cause and consequences: A symposium,* edited by J. Wilbert. Pp. 13–46. Caracas: Sociedad de Ciencias Naturales La Salle.
Lehmann, Walter
    1910   Ergebnisse einer Forschungsreise in Mittelamerika und México, 1907–1909, *Zeitschrift für Ethnologie* **42**, 687–749.
Lewis, Oscar
    1963   *Life in a Mexican village: Tepoztlán restudied.* Urbana: Univ. of Illinois Press.
Long, Edward
    1774   *The history of Jamaica, . . . and account of the Mosquito Shore.* London: T. Lowndes.

Margalef, Ramon
  1968 *Perspectives in ecological theory.* Chicago: Univ. of Chicago Press.
  1969 "On certain unifying principles in ecology." In *Contemporary readings in ecology,* edited by A. Boughey. Pp. 374–390.
Massajoli, Pierleone
  1968 I Sumu e i Mîskito, *Geografico Militare Revista bimestrale dell'Instituto* (Florence), No. 4, pp. 727–780.
  1970 I Miskito: Note Sull 'Acculturazione, *Terra Ameriga* (Italy), No. 20–21, pp. 25–32.
Matthiessen, Peter
  1966 *The cloud forest.* New York: Pyramid.
  1967 To the Miskito Bank. In *The New Yorker,* October 28, 1967, pp. 120–164.
Mauss, Marcel
  1966 *The gift* (Translated by Ian Cunnison). London: Cohen and West.
Meggers, Betty J.
  1971 *Amazonia: Man and culture in a counterfeit paradise.* Chicago: Aldine.
Mikesell, Marvin W.
  1970 Cultural ecology. In *Focus on geography,* edited by Phillip Bacon. Pp. 39–61. Washington: National Council for the Social Sciences.
Mueller, Bishop Karl A.
  1932 *Among Creoles, Miskitos and Sumos.* Bethlehem, Pa.: The Comenius Press.
Nelson, Richard K.
  1969 *Hunters of the northern ice.* Chicago: The Univ. of Chicago Press.
Newman, Marshall T.
  1961 Biological adaptation of man to his environment: Heat, cold, altitude, and nutrition, *Annals of the New York Academy of Sciences* **91**, 617–633.
Newton, Arthur P.
  1914 *The colonizing activities of the English Puritans.* New Haven: Yale Univ. Press.
Nicholls, Lucius
  1961 *Tropical nutrition and dietetics.* 4th ed. London: Baillière, Tindall and Cox.
Nietschmann, Bernard
  1969 The distribution of Miskito, Sumu, and Rama Indians, eastern Nicaragua, *Bulletin of the International Committee on Urgent Anthropological and Ethnological Research,* International Union of Anthropological and Ethnological Sciences (Vienna) **11**, 91–102.
  1971a Destrucción de la Fauna de la Costa Atlántica, *La Prensa,* Managua, Nicaragua, September 5.
  1971b The substance of subsistence. In *Geographic research on Latin America,* edited by B. Lentnek, R.L. Carmin, and T.L. Martinson. Pp. 167–181.
  1972 Hunting and fishing focus among the Miskito Indians, eastern Nicaragua, *Human Ecology* **1**, 41–67.
Nuhn, Helmut
  1969 Landesaufnahme und Entwicklungsplanung im Karibischen Tiefland Zentralamerikas, *Erdkunde,* **23**, 142–154.
Nye, P.H. and D.J. Greenland
  1960 *The soil under shifting cultivation.* Technical Communication No. 51. Commonwealth Bureau of Soils. Harpenden: Commonwealth Agricultural Bureau.
Odum, Eugene P.
  1959 *Fundamentals of ecology,* 2nd ed. Philadelphia: W.B. Saunders.
  1971 *Fundamentals of ecology,* 3rd ed. Philadelphia: W.B. Saunders.
Odum, Howard T. and Robert F. Pigeon
  1970 *A tropical rain forest: A study of irradiation and ecology at El Verde, Puerto Rico.*

268   References

Division of Technical Information, Washington, DC: US Atomic Energy Commission.

Parsons, James J.
1954   English-speaking settlement in the western Caribbean, *Yearbook, Association of Pacific Coast Geographers* **1**, No. 16, 3–16.

1955a The Miskito pine savanna of Nicaragua and Honduras, *Annals of the Association of American Geographers* **45**, 36–63.

1955b Gold mining in the Nicaraguan rain forest, *Yearbook of the Association of Pacific Coast Geographers* **17**, 49–55.

1956   *San Andrés and Providencia, English-speaking islands of the western Caribbean.* Univ. of California Publications in Geography **12**, No. 1, 1–84. Berkeley: Univ. of California Press.

1962   *The green turtle and man.* Gainesville: Univ. of Florida Press.

n.d.   The hawksbill turtle and the tortoise shell trade. France *(in press).*

Parsons, James J. and William M. Denevan
1967   Pre-Columbian ridged fields, *Scientific American* **217**, 92–100.

Pijoán, Michel
1944   The Miskito Indians. Some remarks concerning their health and the lay health program, *América Indígena* **4**, No. 4, 255–263.

1946   The health and the customs of the Miskito Indians of northern Nicaragua: Interrelationships in a medical program, *América Indígena* **6**, No. 1, 41–66, No. 2, 157–183.

Pim, Captain Bedford and Berthold Seemann
1869   *Dottings on the roadside in Panama, Nicaragua, and Mosquita.* London: Chapman and Hall.

Pittier, H.
1892   La parte sureste de la República de Costa Rica, *Anales del Instituto Fisico—Geographico y del Museo Nacional de Costa Rica* **3**, 107–113.

Porter, Philip W.
1965   Environmental potentials and economic opportunities—a background for cultural adaptation, *American Anthropologist* **67**, 409–420.

Portig, W.H.
1965   Central American rainfall, *Geographical Review* **55**, 68–90.

Radell, David R.
1969   "Historical geography of western Nicaragua: The spheres of influence eon, Granada, and Managua, 1519–1965." Report of field work carried out under ONR Contract No. 3656(03), Project NR 388 067. Dept. of Geography, Univ. of California, Berkeley.

Radley, Jeffrey
1960   "The physical geography of the east coast of Nicaragua." Report of field work carried out under ONR Contract 222 (11) NR 388 067. Dept. of Geography, Univ. of California, Berkeley.

Rappaport, Roy A.
1968   *Pigs for the ancestors: Ritual in the ecology of a New Guinea people.* New Haven: Yale Univ. Press.

1971   The flow of energy in an agricultural society, *Scientific American* **224**, No. 3, 116–132.

Richards, Audrey I.
1961(ed) *Land, labour and diet in northern Rhodesia.* International African Institute. London: Oxford Univ. Press (first published in 1939).

Roberts, Orlando W.
  1827   *Narrative of voyages and excursions on the east coast and in the interior of Central America.* Edinburgh: Constable and Co.
Rohrl, Vivian J.
  1970   A nutritional factor in Windigo psychosis, *American Anthropologist* **72**, 97–101.
Roys, R.L.
  1943   *Indian background of colonial Yucatan.* Carnegie Institute of Washington, Publication 548.
Sahlins, Marshall D.
  1965   On the sociology of primitive exchange. In *The Relevance of Models for Social Anthropology,* ASA Monograph 1, edited by M. Banton. Pp. 139–236. London: Travistock publications.
  1968   Notes on the original affluent society. In *Man the hunter,* edited by R. B. Lee and I. Devore. Chicago: Aldine. Pp. 85–89.
  1972   *Stone age economics.* Chicago: Aldine.
Sandner, Gerhard
  1964   La Costa Atlántica de Nicaragua, Costa Rica y Panamá. Su conquista y colonización desde principios de la época colonial, *Informe Semestral,* enero a junio, Instituto Geográfico de Costa Rica, pp. 83–136.
Sapper, Karl
  1900   Reise Auf Dem Río Coco (Nordliches Nicaragua): Besuch Der Sumos Und Mosquitos, *Globus, Braunschweig* **78**, 249–252, 271–276.
Sauer, Carl
  1958   Man in the ecology of tropical America, *Proceedings Ninth Pacific Science Congress,* (1957), **20**, 104–110. Bangkok: Secretariat, Ninth Pacific Science Congress, Dept. of Sciences.
Sloane, Hans
  1709   *A voyage to the islands Madera, Barbadoes, Nieves, S. Christophers and Jamaica. . . .* 2 Vol. London.
Squier, Ephraim George (Bard, Samuel A. is pseudonym)
  1855   *Waikna: Adventures on the Mosquito Shore.* New York.
  1858   *The states of Central America.* New York.
Stefansson, V.
  1960   Food and food habits in Alaska and northern Canada. In *Human nutrition: Historic and scientific,* edited by I. Galdston. Pp. 23–60. New York: International Univ. Press.
Stoddart, David R.
  1962   Catastrophic storm effects on the British Honduras reefs and cays, *Nature* **196**, 512–515.
  1963   *Effects of Hurricane Hattie on the British Honduras reefs and cays, October 30-31, 1961.* Atoll Research Bulletin No. 95. Washington: Pacific Science Board and National Academy of Sciences-National Research Council.
  1965   Re-survey of hurricane effects on the British Honduras reefs and cays, *Nature* **207**, 589–592.
  1969   *Post-hurricane changes on the British Honduras reefs and cays: Re-survey of 1965.* Atoll Research Bulletin No. 131. Washington: Pacific Science Board and National Academy of Sciences-National Research Council.
Stone, Doris
  1966   Synthesis of lower Central American ethnohistory. In *Handbook of Middle American Indians,* edited by R. Wauchope. Vol. 4, pp. 209–233. Austin: Univ. of Texas Press.

Storr, John F.
  1964   *Ecology and oceanography of the coral-reef tract, Abaco Island, Bahamas.* Geological
          Society of America Papers, No. 79.
Strangeways, Thomas
  1822   *Sketch of the Mosquito Shore, including the territory of the Poyais.* Edinburgh.
Street, John M.
  1969   An evaluation of the concept of carrying capacity, *The Professional Geographer*
          21, No. 2, 104–107.
Swift, Raymond W.
  1959   Food energy. In *Food, the yearbook of agriculture 1959*, edited by Stefferud. Pp.
          39–56. Washington, DC: US Dept. of Agriculture.
Taylor, B.W.
  1959   *Ecological land use surveys in Nicaragua.* Vol. 1 Managua: Ministerio de Economia,
          Instituto de Formento Nacional.
  1963   An outline of the vegetation of Nicaragua, *Journal of Ecology* **51**, 27–54.
U.S. Hydrographic Office
  1948   *Weather summary of Central America for use with naval air photos.* US Hydrographic
          Office Publication 531. Washington: Dept. of Commerce.
Vayda, Andrew P. and Roy A. Rappaport
  1968   Ecology: Cultural and non-cultural. In *Introduction to cultural anthropology,* edited
          by J.A. Clifton. Pp. 477–497. Boston: Houghton Mifflin.
Vermeer, Donald Eugene
  1963   Effects of Hurricane Hattie, 1961, on the cays of British Honduras, *Zeitschrift
          für Geomorphologie* **7**, 332–354.
  1964   "Agricultural and dietary practices among the Tiv, Ibo, and Birom tribes,
          Nigeria." Unpublished Ph.D. dissertation. Dept. of Geography, Univ. of Califor-
          nia, Berkeley.
Vivó Escoto, Jorgé A.
  1964   Weather and climate of Mexico and Central America. In *Handbook of Middle
          American Indians,* edited by R.C. West, Vol. 1, pp. 187–215. Austin: Univ. of
          Texas Press.
Von Hagen, V.W.
  1940   The Mosquito Coast of Honduras and its inhabitants, *Geographical Review* **30**,
          238–259.
W., M.
  1732   The Mosqueto Indian and his golden river; being a familiar description of the
          Mosqueto Kingdom of America (written about 1699.) In *A Collection of Voyages
          and Travels,* edited by A. Churchill. Vol. 6, pp. 297–312. London.
Waddel, Eric
  1972   *The mound builders.* Seattle: Univ. of Washington Press.
Weiss, Brian
  1971   "The ecology of readaptation." Unpublished paper, Dept. of Anthropology,
          Univ. of Michigan.
West, Robert C.
  1964   Surface configuration and associated geology of Middle America. In *Handbook
          of Middle American Indians,* edited by R.C. West. Vol. 1, pp. 33–83, Austin: Univ.
          of Texas Press.
West, Robert C. and John P. Augelli
  1966   *Middle America: Its lands and peoples.* Englewood Cliffs: Prentice-Hall.

Whiting, M.G.
1958 "A cross-cultural nutrition survey of 118 societies representing the major cultural areas of the world." Unpublished Ph.D. dissertation. Harvard School of Public Health, Harvard.

Wickham, Henry A.
1872 "A journey among Woolwa or Soumoo Indians of Central America." Part 2 of *Rough Notes of America Travel*, pp. 143–287 London: W.H.J. Carter.

Williams, William
1969 *Mr. Penrose: The journal of Penrose, seaman.* Edited by D.H. Dickason. Indiana Univ. Press.

Woodburn, James
1968 An introduction to Hadza ecology. In *Man the hunter*, edited by R.B. Lee and I. Devore. Pp. 49–55. Chicago: Aldine.

Wu Leung, Woot-Tsuen
1961 *Food composition table for use in Latin America*. A research project sponsored jointly by INCAP-ICNND: Bethesda, Maryland: National Institutes of Health.

Young, Thomas
1847 *Narrative of a residence on the Mosquito Shore with an account of Truxillo, and the adjacent islands of Bonacca and Roaton.* London.

INDEX

X